Green Energy and Technology

Climate change, environmental impact and the limited natural resources urge scientific research and novel technical solutions. The monograph series Green Energy and Technology serves as a publishing platform for scientific and technological approaches to "green"—i.e. environmentally friendly and sustainable—technologies. While a focus lies on energy and power supply, it also covers "green" solutions in industrial engineering and engineering design. Green Energy and Technology addresses researchers, advanced students, technical consultants as well as decision makers in industries and politics. Hence, the level of presentation spans from instructional to highly technical.

More information about this series at http://www.springer.com/series/8059

Vinod Kumar Verma · Karel Katovsky

Spent Nuclear Fuel and Accelerator-Driven Subcritical Systems

Springer

Vinod Kumar Verma
University of Rajasthan
Jaipur, Rajasthan, India

and

GGSIP University
New Delhi, India

Karel Katovsky
Department of Electrical Power Engineering, Faculty of Electrical Engineering and Communication
Brno University of Technology
Brno, Czech Republic

ISSN 1865-3529 ISSN 1865-3537 (electronic)
Green Energy and Technology
ISBN 978-981-10-7502-5 ISBN 978-981-10-7503-2 (eBook)
https://doi.org/10.1007/978-981-10-7503-2

Library of Congress Control Number: 2018950201

This Springer imprint is published by the registered company Springer Nature Singapore Pte Ltd.
The registered company address is: 152 Beach Road, #21-01/04 Gateway East, Singapore 189721, Singapore

Preface

Spent nuclear fuel (SNF) is a fallout of nuclear energy programs of the present-day society. In fact, disposal of the radioactive waste (RAW) is a challenge and every human is threatened by it.

It reminds us that thousands of years back the human race was frightened of volcanoes and electric thunders but they provided the concept of the source of fire and hot material. The human race took it as a challenge and spent thousands of years in developing different ways of producing fire and finding thousands of applications too. On the social front, the development of fire even gave birth to cultural and racial problem in different parts of the globe. But, after all, it provided the means of positive developments and showed a definite path of advancement to the entire human race of the time. It may be assumed that in the modern time, the human race is attempting to handle the new, compact, and wonderful source of heat and energy in the form of nuclear energy.

Every energy issue has raised eyebrows of the strong believers of humanity because for the infinite time the sun has been the source of energy and life with which the human race has its strongest relationship. The sun being at far long distance does not raise any intimate threat to the mankind. On the other hand, SNF has a relatively intimate relationship. Its heaped amount on earth and having a feature of its reutilization cannot allow it to be avoided from the developmental process. The book is written for all these concerns so that it can help our education system and future developments.

The nuclear science community is well aware of both threats and applications of the nuclear energy and it is highly concerned with finding solutions. RAW contains a fatal radio-toxic radiation and at the same time a bigger proportion of RAW can be reutilized as a fuel also. It is the result of serious concerns only that despite of the fact that hundreds of nuclear energy reactors are operational world over and there have been only a few accidents. Small accidents can be avoided by generating awareness through education and developing stronger control systems regarding proliferation and sources of threats. This is also deeply related to education that nuclear energy has been grown for peaceful utilization.

While producing the text of the first chapter of the book, we thought several times that how the issue of SNF is a state-of-the-art issue. Ultimately, we kept the subtitle on having a keen look at the national ways of handling the international issues, treaties and standards in their own references, national interests and sometimes agreeing on an issue and other times disagreeing. These details are presented at the end of the chapter. Still, it can be emphasized that proper education can solve these petty issues in future.

The writing of the book started with the fact that over the last twenty years or more the authors have participated in several experiments and developmental works, organization of several meetings and conferences and producing the manpower for both research and education, and a book is considered to be the best carrier of compilation of such knowledge. Several books have been written on the issues of proliferation and nuclear waste but only a few articles or the periodicals are able to inspire the young minds to begin with a 'start up' and career in research. The book will arouse that kind of keenness for both. The book will support the graduate students and innovators to grow new ideas of research and entrepreneurship also, in the field of new setup of energy utilization and to save from the energy losses.

Efforts have been made that the book comprises existing knowledge till last few years for education purpose. The readers may provide help to the authors in pointing out more details that may be added in the future version of the book. It may, however, be pointed out that the chapter related to data is kept short as one of the authors had presented many recent data in his earlier book entitled *Role of (n, xn) reactions in ADSS* published in the year 2011.

The authors have a deep sense of gratitude to their co-workers who had participated in different experiments with them because they had helped in developing the concept without even known to us about writing of the book. The authors are grateful to the international community who is involved in various related research and development activities and they helped us in providing some of the results of their hard work in response to our personal contacts. The authors are grateful to JINR, Dubna where they used its facilities and entire research framework for their career.

Our sincere thanks are due to Dr. N. S. Raghaw for his help in providing the data of radiation damage, handling various data files and data sites and running the JA-IPU code even at a call at an odd hour. His enormous help at the time of final compilation of the manuscript was commendable.

The authors are thankful to Springer for having the patience for the completion of the manuscript and their prompt responses all throughout its preparation for over a year.

Jaipur/New Delhi, India — Dr. Vinod Kumar Verma (Superannuated Professor)

Brno, Czech Republic — Prof. Karel Katovsky

February 2018

Contents

About the Authors

Dr. Vinod Kumar Verma was Professor of physics at the University of Rajasthan, Jaipur, India, until April 2010 and Professor of engineering physics/nanoscience at the GGSIP University, New Delhi, until January 2016. In his research career spanning more than 40 years, he participated in international collaborations like EMU01, L3, WA93, WA98, GAMMA-2 and 3, E+T, and QUINTA at the Joint Institute for Nuclear Research (JINR), Dubna. He has authored and co-authored over 180 research papers and edited the proceedings of the Workshop on Physics of Accelerator Driven Sub-critical Systems (ADSS). In India, he is best known for his pioneering research on 'spallation neutrons' in the course of several S&T projects completed for DST and DAE (Government of India). He has also successfully completed developmental work on two Monte Carlo codes for the ADSS technology, i.e. the CASCADE 2004 version in collaboration with JINR, Dubna and the JA-IPU code for radiation damage with a team of Indian researchers. Over the past 20 years, he has initiated scientific collaborations with teams in the Czech Republic, S. Korea, and Russia on the activities related to ADSS technique. Present book is his second contribution for education and research activity on the ADSS.

Dr. Karel Katovsky graduated with a degree in nuclear engineering. His Ph.D. thesis at the Faculty of Nuclear Sciences and Physical Engineering of the Czech Technical University in Prague (2008) was on the experimental research and the nuclear data problems of ADSS. He subsequently spent several years at JINR, Dubna. He worked at the Computation Reactor Physics Department of SKODA JS, where he dealt with nuclear fuel loading patterns and their optimization. Since 2012, he has been working as Assistant Professor of nuclear engineering at the Brno University of Technology, where he serves as Leader of the Nuclear Power Group to teach nuclear energy. He is the author or co-author of 37 SCOPUS-indexed papers, more than 50 conference papers, and about 30 research reports. He is a member of the Board of the Czech Nuclear Society and of the expert committee for nuclear energy of regional NGO associations in the NPP Dukovany area.

Spent Nuclear Fuel and Alternative Methods of Transmutation

1

The chapter deals with the status of the spent nuclear fuel and accumulation of unspent nuclear fuel world over. Its fertile, fissile and fission product components are also discussed along with their applications and various methods of reprocessing the SNF. International situation related to reprocessing, security aspects arising from the SNF, and the fissile components along with its handling at individual national level are also summarized.

1.1 Spent and Unspent Nuclear Fuel and the Nuclear Waste

Spent nuclear fuel (SNF) is the nuclear fuel that is irradiated in a power reactor. A few percent of this is utilized in power generation, and a very large part is left behind as a radiotoxic material at the time of discharge of a reactor. In the SNF major actinides, U and Pu are 95–96 and 1%, respectively. Minor actinides (Np, Am, and Cm) are 0.1%, short-lived fission products (FP) are 3–4%, and long-lived FPs are 0.1%. Different composition elements pose different challenges for disposition of SNF. In fact, unspent nuclear fuel (UNF) is of high concern because its components can be utilized for nuclear energy or other atomic devices with least efforts.

1.1.1 State-of-the-Art: Spent Nuclear Fuel Issue

1.1.1.1 Definition of the Problem

By definition, UNF is the fuel that is left after reprocessing of the radiotoxic nuclear material which is discharged from a reactor and that can qualify for re-application in another reactor. Reprocessing is a combined physical and chemical process that acts on the basis of the kind of radioactive material discharged from a reactor, for example, actinides and fission products. Actinides may further be classified as isotopes of uranium or plutonium, fertile or fissile, etc. A nuclear power reactor that produces 1 gigawatt electric power (1 GWe) burns annually $\sim$1 ton of fissile fuel which is equivalent to 2 million ton of oil equivalent (toe). According to an inventory [1], a reactor loaded with 26,328 kg of ^{238}U and 954 kg of ^{235}U gives out 280 kg of ^{235}U, 111 kg of ^{236}U, 266 kg of Plutonium (total), 946 kg of total fission products (FP), and 25,655 kg of ^{238}U at the time of discharge of the reactor. As a matter of fact, a nuclear reactor produces enormous amount of energy which is a clean energy compared to that from any fossil fuels like oil and coal used in energy production and they have threatened the environment seriously in last 50 years. Production of radiotoxic materials in

V. K. Verma and K. Katovsky, *Spent Nuclear Fuel and Accelerator-Driven Subcritical Systems*, Green Energy and Technology, https://doi.org/10.1007/978-981-10-7503-2_1

the form of FP and minor and higher actinides in a nuclear power reactor are otherwise manageable, and because of strong security acts, it takes several years to convert it to the unspent nuclear fuel. It is dumped in a repository most of the time.

1.1.1.2 Amounts in Different Reactor Types

Unspent nuclear fuel inventory at time of discharge depends on the loaded fuel inventory, design of a reactor from the point of neutron spectrum, burning of the fuel, production of actinides, hence the fuel cycle and the conditions of shut down or discharge of the reactor. From the point of design, a normal power reactor is different to a breeder reactor. Also, there can be differences from the point of a light water or a heavy water cooled reactor. In most of the reactors, UO_2 fuel cycle is a popularly used fuel cycle. But, there are possibilities of enrichment of UO_2 or even mixing of Plutonium or another fissile fuel such as PuO_2–UO_2 fuel and that has impact on discharge inventory [2]. There can be several possible inventories of loading of a reactor and one of the PFBR [3] with mixed uranium–plutonium MOX fuel, a mixture of PuO_2 and UO_2 in the form of several isotopes that are organized in the two-core reactor is shown by the data in Table 1.1. They are generally categorized as the fast reactors. In fact, fuel mixture and design of loading for a reactor is state-of-the-art technique.

From the highly radioactive discharged material of a reactor which has more than 95% unspent original fuel, newly formed fissile fuel and a large number of elements that find applications in medical science, industry and other applications, extraction of useful material is done several times, albeit, a highly costly affair. So long as extraction is done or the discharged material is planned to be dumped as such it is treated as a nuclear waste.

1.1.1.3 Issue of Radioactive Waste Management

Nuclear waste comprises of liquids and chemicals of a processing plant, scrapes from milling and leftover nuclides after medical applications including the files and containers or the canisters used in transportation as well as the sand and concrete of the ground of processing units. Nuclear waste is classified as (i) low-level waste (LLW), (ii) intermediate level waste (ILW), (iii) high-level waste (HLW), and (iv) trans-uranic-level nuclear waste (TRUW) according to the radio-toxicity which is discussed in next Sect. 1.1.2 and is subjected to dump, reposition, or other means of dissemination. In some quarters of the technology, unprocessed disposal of a reactor is also named as the 'used nuclear fuel' because of existence of a possibility of application of a part of this as a fuel after extraction. According to one report [4] in US alone nearly 60,000 MT (metric ton) of UNF existed up to the year 2009, and it will be discussed in detail in later part of the chapter.

1.1.2 Radio-Toxicity

Radio-toxicity is always measured by the quantity of radioactive material or the radioactivity alone and in case of biological hazard it has nothing to do with chemical toxicity of the hazard [4]. Ingested or inhaled radio-toxicity is commonly expressed in 'Sieverts (Sv)*'. The radio-toxicity of a given radioactive isotope is equal to the product of multiplication of its activity (Bq) and effective dose coefficient $e(\tau)$ for ingestion or similarly, for inhalation for a period τ expressed in years. The effective dose coefficient $e(\tau)$ (units Sv/Bq) is defined as the committed effective dose per unit acute intake for a period, τ. Equation (1.1) shows the relationship

Table 1.1 Minor actinide (%) in core 1 and core 2 subassembly of a PFBR reactor

Core	Minor actinide (%)					
Core 1/core 2	13.78/18.48	13.94/18.64	14.4/19.6	15/20.3	15.6/21	16/22. 2

between radio-toxicity (Sv/kg) and the effective dose coefficient $e(\tau)$ [4].

$$\text{Radio}-\text{toxicity (Sv)} = \text{Activity (Bq)} \times e(\tau)\,(\text{Sv/Bq}) \quad (1.1)$$

Radio-toxicity is also expressed as volume of water or air in units of m^3 required to dilute the radioactive material potential below the threshold limit for the general public. In case of ingestion, it is the volume of 'water in m^3' and in case of inhalation, it is the volume of 'air in m^3'. The threshold limit for general public is 50 mrem/ year or equivalently 0.5 mSv/year based on average intake of 2 L/day of water or 22 m^3/day of air. Radio-toxicity, expressed in Sv/kg of some of the elements of UNF, is given in Table 1.2 for calculation of total radio-toxicity for a given amount of radionuclide in kg. For details of effective dose coefficient, $e(\tau)$ of radionuclides refer to updated ICRP 119 [5] in place of ICRP 72.

Different radiations like gamma, electron, muon, neutron, proton, and heavier charged particle give different doses to different tissues and organs. A body has several tissues and organs. If H_T is defined as the equivalent dose in a tissue or an organ T, then the sum of weighted equivalent doses in all tissues or organs of the body can be written as $E = \text{Sum}\ (W_T H_T)$. Here, W_T is the weight factor for a tissue. For the radiation dependence of equivalent dose, $H_{T,R} = W_R\ D_{T,R.}$ Here, $D_{T,R}$ expressed in (J/kg) is the average absorbed dose for a radiation R in a tissue T. Thus, $H_T = \text{Sum}\ H_{T,R}$. Weighting factor W_R signifies quality of a radiation being smallest for gamma and higher for different energy neutrons. For proton and heavier charged particles, W_R is even higher. Values of W_R as per ICRP 119 for different radiation are given in Table 1.3. It can be understood from the data that alpha-emitting radionuclides have higher values of $e(\tau)$ than β-emitters. Different energy neutrons have different values of W_R** due to different possibilities of activations. All alpha and heavier mass charged particles are kept in one category.

1.1.3 Extraction or Reprocessing

Extraction of actinides or fission product (FP) from the discharged spent fuel is being processed since 1940 using precipitation process. Adopting a route for reprocessing is based on the kind of a reactor. Following three technologies are developed by various countries,

a. Hydrometallurgy
b. Pyrometallurgy
c. Fluoride volatility

Hydrometallurgy process is rather one of the oldest and a matured technology used for separation of major actinides like plutonium and uranium as well as for the conditioning of the ultimate waste for long-term storage. As the name suggests, it is aqueous and in this category following methods are popular.

(i) Standard PUREX (**P**lutonium **U**ranium **Re**dox **Ex**traction) and extended PUREX
(ii) UREX
(iii) COEX
(iv) NUEX

First PUREX plant using the PUREX process was opened up at Savannah River in 1954. Several reprocessing plants started operation in Belgium, France, Germany, India, Japan, Russia, and UK during 1960–1970. Methods UREX, COEX, and NUEX are the derivations of basic PUREX method [6] and differ in producing not the pure Plutonium.

According to the PUREX process, after nearly 5 years of cooling period of a discharge, in order to dissolve the spent fuel it is exposed to nitric acid solution to give nitrates of plutonium, uranium, minor actinides (Np, Am, and Cm), and the fission products. Volatile elements like iodine, krypton and xenon are removed for the off-gas treatment. More noble fission products are not dissolved and they are treated with care along with the fuel assemblies. The solution with dissolved species is forwarded to the extraction process that depends on affinities with the

Table 1.2 Data of radio-toxicity [1] of some of the long-lived actinides and fission products

Element	Half-life (year)	Decay mode	Radioactivity (Bq/kg)	$e(\tau)$ (Radio-toxicity)
$^{236}Np_{93}$	1.54×10^5	ε (87.3), β (12.5), α (0.16)%	–	–
$^{237}Np_{93}$	2.14×10^6	α	2.6×10^{10}	110 (0.3×10^4)
$^{238}Pu_{94}$	87.7	α	6.3×10^{14}	230 (1.4×10^8)
$^{239}Pu_{94}$	2.41×10^4	α	2.3×10^{12}	250 (0.6×10^6)
$^{240}Pu_{94}$	6.564×10^3	α	–	250
$^{241}Am_{95}$	4.33×10^2	α	1.3×10^{14}	200 (0.3×10^8)
$^{242m}Am_{95}$	141	α (0.46%)	–	–
$^{243}Am_{95}$	7.37×10^3	α	–	200
$^{244}Cm_{96}$	18.1	α	–	120
$^{245}Cm_{96}$	8.5×10^3	α	–	210
^{79}**Se**	3.27×10^5	β	–	–
^{93}Zr	1.53×10^6	$\beta\gamma$	–	–
^{97}Tc	4.21×10^6	ε	–	–
^{98}Tc	4.20×10^6	β	–	–
^{99}Tc	2.11×10^5	β	6.3×10^{11}	(4.9×10^2)
^{107}Pd	6.5×10^6	β	–	–
^{126}Sn	1.0×10^5	β	–	–
^{129}I	1.57×10^7	β	6.5×10^9	(0.7×10^3)
^{135}Cs	2.3×10^6	β	–	–

*Named after Rolf Maximillan Sievert, 1 Sv = 100 rem = 1 J/kg for absorbed dose. Also, expressed in $m^2\ s^{-2}$ in MKS. Sievert is used in connection of a biological component and Gray (J/kg) for any non-biological physical object.
**Tissue weighting factor is used as radiation weighting factor, W_R to represent relative biological effectiveness according to ICRP 119. This depends on the kind of a radiation. See Table 1.3 for W_R for different radiation. For an individual organs W_T are used

Table 1.3 Radiation weighting factor W_R as per ICRP 119 publication [6]

Radiation weighting factor		
S. No.	Type of radiation and energy range	W_R
1	Photons of all energies	1
2	Electrons and muons all energies	1
3	Neutrons, energy (i) <10 keV (ii) 10–100 keV (iii) >100 keV–2 MeV (iv) >2 MeV–20 MeV (v) >20 MeV	5 10 20 10 5
4	Protons, other than recoil, energy >2 MeV	5
5	Alpha particles, fission fragments, heavy nuclei	20

aqueous or organic phases. As a result of this, FP and minor actinides are cooperatively on one side and uranium and plutonium on other side. On a more specific electron state manipulation, plutonium remains with the organic phase and uranium is separated on making aqueous phase which is less acidic. After separation, they are solidified through the de-nitration in case of uranium and oxalate precipitation and calcinations in case of plutonium for the purpose of storage. The solidified uranium and plutonium are stored in separate tanks, and the leftover solvent still has fission products and the actinides which may further be extracted. In PUREX extraction, nearly 99.87% of uranium and 99.36–99.51% of plutonium are extracted. Plutonium is obtained as an oxide in the form of a powder. The PUREX is subject to several risks of proliferation and transportation. Uranium and plutonium so extracted can be converted to MOX fuel by mixing them.

UREX is a solvent extraction process for extracting uranium and technetium (Tc) from the SNF leaving the plutonium and FP behind in the solution. In UREX method because plutonium is not extracted, it is treated as proliferation resistant. In the process, 99% U and 95% Tc are recovered and rejecting of 99% of trans-uranic isotopes.

In COEX, nuclear fuel is extracted in three streams (i) uranium–plutonium, (ii) uranium, and (iii) fission products and minor actinides, but never plutonium alone for avoiding the proliferation risk. From the first stream, extracted U-Pu is converted into MOX (mixed oxide) fuel for the light water reactors (LWR). The uranium extracted in second stream is sent for several subsidiary processes such as purification, conversion, enrichment, and building as a fuel for a reactor. The fission products and minor actinides are sent for vitrification into the glass logs. Vitrification is a process of capsuling small amount of radioactive material in small glass bids for reposition.

NUEX is a fourth-generation aqueous-based reprocessing system. The process works as first shearing and then dissolution of SNF in nitric acid. Uranium, plutonium, and neptunium are then extracted into a solvent tr-butyl phosphate made by dissolving in odorless kerosene. In primary separation, uranium stream and mixed uranium–plutonium and then neptunium streams are produced. Later, bulk uranium is separated from technetium. Mixed stream is purified from ruthenium. The technique is well established and promoted in France. The conceptual layout of NUEX can be shown as follows Fig. 1.1.

Commissioning of a reprocessing plant is a task involving high risk because of a large number of issues. Some of them can be briefly mentioned as follows,

(i) high cost
(ii) long design, construction commissioning period
(iii) international reprocessing services become even more difficult by way of rules related to safe transport of nuclear fuel, return of all products and the waste with proper packing and certification holding the radiological safety protocols.

Russia, France, and UK have offered reprocessing services internationally and the contracts are offered with full international safe guards. In last two decades, interest has been shown in

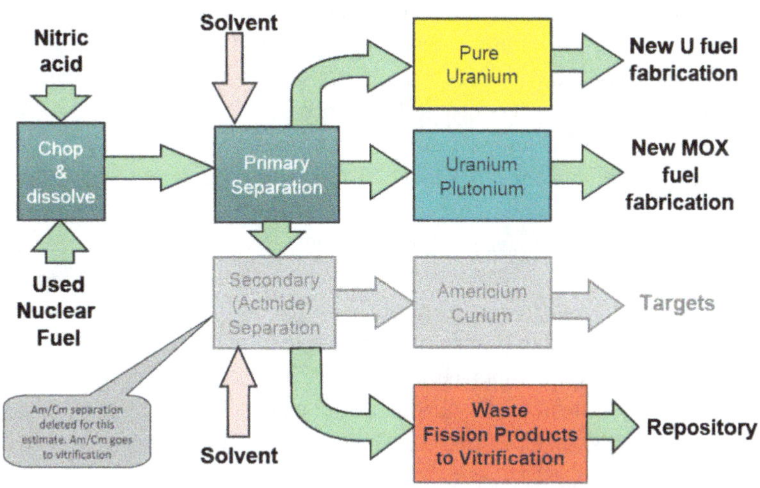

Fig. 1.1 Conceptual design of NUEX processing [7]

more innovative technologies of spent fuel treatment having chance to emerge as 'alternative method' to the conventional PUREX processes. These technologies may deserve attention for developing advanced systems by dry technologies, among others,

- integral fast reactor concept for metallic fuel recycle
- vibropac method for oxide fuel recycle
- DUPIC concept for recycle of spent fuel of LWR and HWR.

Metallic fuel of a fast reactor is processed by electrolytic processes. LiCl-KCl with added $CdCl_2$ is used as the electrolyte which is filled in the steel basket and uranium from the cut pieces of fuel pins is transported to the cathode at a rate of 3 g per ampere-hour in the electrolysis process. The $CdCl_2$ helps in converting some of the fuel elements into their chlorides. Trans-uranic metals are separated using a setup of different electrodes. 'Pyro-processing,' uses electrical current to sift out the useful fuel elements and does not separate pure plutonium. Uranium is transported to the solid cathode containing small amount of electrolyte salts. In fact, at the cadmium cathode which is suspended in the electrolyte salt, mixture of Pu, Am, Np, Cm, U, and the rare earth FPs are collected [6, 8]. Other FPs are left behind for another processing.

France produces more than 75% of its electricity requirement from nuclear energy, and it also reprocesses the spent fuel for itself as well as for foreign countries in its three reprocessing plants UP1, UP2 (based on PUREX route), and UP3. According to reports [9–12], it has processed 18,000 tHM (tHM = ton of heavy metal) of graphite gas-cooled reactor (GCR) and 22,700 tHM of LWR for itself and other countries. Their capacity of reprocessing is $\sim$1,700 t/year including processing of heavy metals. As a matter of safety, at one of the reprocessing plant at La Hague average occupational exposure has been reported to be 0.073 mSv/year/employee, below the natural background despite of the fact that La Hague plant fabricates the MOX fuel assemblies. This is far less than many other plants working in Europe. Also, at La Hague the surrounding population has added a dose of 0.01 mSv/yr/person only over a long period of 20 years. As the overall averages are very small, it may be assumed that chance of accumulation of high dose on a group of employees is far away to reach to the critical limit.

Presently, 16 power reactors are operating in India with total power capacity 3.9 GW. Out of them, 2 are BWRs and remaining 14 are PHWR. India has an integral three-stage energy program [13] of reprocessing and recycling of the fuel components for an appropriate reactor. In its second stage, fast breeder reactors (FBRs) are to be backed by the reprocessing plants and plutonium-based fuel fabrication plants. India had its first PUREX reprocessing plant at Trombay for its research reactor and later in 1975, another plant was raised at Tarapur for reprocessing spent fuel of its PHWRs. Its third reprocessing plant was commissioned at Kalpakkam in the year 1998 for meeting the requirements of its nuclear energy program after including the fast breeder reactors. The extracted fissile fuel will be utilized to meet its energy need from second phase onward. For utilization of thorium, Advanced Heavy Water Reactors (AHWR) will employee both $(Th–Pu)O_2$ and $(Th\text{-}^{233}U)O_2$ fuels as one cluster in third stage of its nuclear energy program. This will introduce a challenging task of reprocessing of three components U, Pu, and Th. Most of the experience in THOREX domain has come from the recovery of low amounts of ^{233}U bred in irradiated ThO_2 in CIRUS research reactor. An engineering scale facility is in operation at Trombay for the processing and recovery of ^{233}U from CIRUS and Dhruva irradiated thorium fuel rods. Studies are being carried out to extract from the ThO_2 fuel bundles of PHWRs.

For some specific details, one can refer to [14, 15] for reprocessing in India.

Tokai processing plant was commissioned in 1981. This can process even MOX spent fuel. Japan through its PUREX reprocessing method has a capacity of processing 800 t U/year and has developed a SNF storage capacity of 3,000 t.

Table 1.4 Status of SNF, reprocessing capacity, direct disposal, and leftover spent fuel in the nuclear energy countries

Country	SNF in year 2005 including MOX (projected with year)	Reprocessing capacity and projected for year 2020 or in future	SNF in year 2007 (tHM)	Pu-in stock
USA	194 tHM (civil). More than 70,000 t is in storage [16]	Reprocessed so far 194 t [11]. 2000 t/yr subject to future permissions	61,000	250–500 t
France	40,700 tHM	~1700 t/yr [11]; 18,000 GCR + 18,000 tHM of LWR is reprocessed [11]	13,500	
Canada		Direct disposal	38,400	
Finland		Direct disposal	1,600	
Germany		Direct disposal now	5,850	
Sweden		Direct disposal	5,400	
UK		2400 t/yr, 3 facilities [11]	5,850	
Belgium	105 tHM	No Data		
Russia	67,000 tHM (Ref. Fig. 1 [12])	400 t/year [11]; (2500 t/year, in year 2020 [11])	13,000	
South Korea		Storage or disposal	10,900	
Japan	1018 tHM of max. 3000 tHM capacity	890 t/yr [11]	19,000	
India	No data	260 t/yr (being upgraded to 560 t/yr) [11]	No data	Civil, 2.9 t
China	No data	(825 t/yr) [11]	60–80 tHM/yr projected [11]	
SNF as data	**1,79,017 tHM**	**6669 t/year capacity**	**~1,74,500 tHM + China**	
Gross SNF available in either form from different projections	2,68,000 tHM (2004) [10]. 3,40,000 tHM (2010) [10]. 4,55,000 tHM (2020) [8]	Total reprocessed 90,000 tHM [10]		

A blank column means no data is available. In the last row, projected figures are given which corresponds to commercial and military sources of SNF. In case of countries like Russia and USA, it may be assumed that big amount of SNF is sent to repositories

In Russia, rate of accumulation of SNF from its own VVER-1000 and RBMK-1000 reactors and from the imported fuel was 600 tHM/year in the year 2006, and it will rise to 1,000 tHM/year in the year 2020. Although Russia has big dry storage capacity, yet believes in reprocessing. Reprocessing capacity of one of its RT-1 plant which works on 'modified PUREX' route is 400 t/year. So far, it has reprocessed ~4,000 t uranium at its Mayak plant. Looking at the accumulation of thermal SNF and plutonium load requirement of new fast reactors, Russia is constructing a new reprocessing plant of capacity ~1,000 t/year [10].

For detailed discussion of chemistry related to PUREX method, readers are advised to refer to [11].

1.1.4 Status of the Unspent Fuel

Unspent fuel can be defined as the fuel component of reprocessed spent fuel. Naturally, this depends on the reprocessing capacity. In estimation of amount of UNF, new fuel accumulated from the mining is not included for the time being. In Table 1.4, an account of the spent nuclear fuel, processing capacity, and amount of the unspent nuclear fuel is given from the resources, some of them are already explained in the previous section.

Reprocessing has been an alternative for managing the spent fuel of a reactor or reactors and other nuclear installations. According to an IAEA report [11], by utilizing the PUREX route about one-third of the SNF has been reprocessed and remaining is stored as inventory either in conventional pools or in the more recently developed dry storage systems. After the year 2005, the gap between amount of spent fuel and reprocessed fuel has increased. In the year 2015, the amount of unspent nuclear fuel was ~25% only, i.e., out of 400,000 metric ton of spent fuel only ~100,000 tons is processed. In Fig. 1.1 of Ref. [11], detailed statistics of the two data of spent and reprocessed fuels from the year 1990 to future projection up to the year 2020 is given. This shows a regular enhancement of gap between amount of SNF and the reprocessed one. In the year 2005, alone in commercial reactors of USA, nearly 2,000 t/year spent fuel is processed and about 50,000 t out of total 70,000 t was in surface storage. As a matter of policy, since 1977 they prohibited reprocessing and preferred to bury it. This is still continued despite of the fact that France, UK and Japan decided to reprocess and to use the unspent mixed oxide fuel for the reactors. It is also true that none of the reactors have so far used the so obtained mixed oxide fuel. It is a hard fact that the unspent nuclear fuel from reprocessing is much costly and chances of proliferations are also not nil. This may also be worth to note that public opinion is that 'why a valuable chunk of wealth is dumped?' and as soon as reprocessing in USA is accepted then the nuclear industry will have large number of employment and energy crises will no more be there. Almost similar situation is in Russia where reprocessing is not promoted probably because there is no crises of fresh nuclear fuel. According to second view of union concerned scientists (UCS) that reprocessing will further enhance already stored piles of plutonium in USA from 250 to 500 t. One reprocessing facility under construction in USA has the objective of converting the surplus plutonium of military stockpiles into mixed oxide fuel for utilization to produce electricity is a welcome step. According to one report of IPFM prepared in the year 2013 [17], total 495 ± 10 tons separated plutonium is in existence globally up to the year 2012.

We know that from a LWR, nearly 2% fissile fuel is retrieved. Nearly half of it is the plutonium. Assuming total spent fuel which is left unprocessed as 175,000 t, there should be 3,500 t of fissile in the year 2007. Of this, there should be 1,705 t of plutonium.

It may also be pointed out that according to IAEA report [11], several of the reprocessing facilities having capacity >1 t/yr have been decommissioned up to the year 2000 in the countries like France (1993) and India (1977) and work on several others was continued. In India, the Trombay facility is made re-operational in the year 1983.

1.1.5 Economy of Reprocessing

In USA, one of the important issues regarding cost comparison of direct dump and reprocessing is dependent on the kind of reactor producing the SNF. Also, there is a concern whether the reprocessing is done first time for second reactor and then to put the disposal of second reactor in a repository. Thermal reactor uses enriched uranium and fast reactors use plutonium enrichment as source of energy. As a result, fast reactors produce more Pu from ^{238}U. Obviously, reprocessing becomes different. The issue becomes more complex from the point of reposition of minor actinides and the FP. According to Congressional Budget Office (USA) [18] cost estimate would incorporate the following,

(i) Cost of handling spent fuel after its discharge from a reactor: Service cost of recovering uranium and plutonium from SNF and to fabricate this into usable fuel assemblies, transport, long-term disposal of the waste are not small and need serious considerations. This needs to be compared with the cost of equivalent new fuel to be purchased for the reactor in absence of reprocessed one.

(ii) In case of direct disposal, cost of interim storage, cooling for 3–5 years, transportation, and cost of long-term reposition is normally counted. In case that the SNF is to be retrieved back and processed after few years, then extra cost of transport and hiked rates will add to the cost of disposal.

Dump of a wealth without its use at least one time will always remain a big question.

According to an estimate by Boston Consulting Group, cost of reprocessing is 585 USD per kg and cost of direct disposal is 555 USD per kg. The Kennedy studies [19] include the issues related to cost of trees equivalent to the volume of the disposal and submit the cost of reprocessing to be 700 USD more than the direct disposal. None of the two included the cost difference for the situation if dumped SNF is to be reprocessed after several years of the dump. After several other considerations explained in Ref. [18], BCO on taking cost data developed by DOE calculated that the repository cost is 1036 USD per kg and it is much higher than the cost estimates submitted by BCG and Kennedy School of Government.

It may be pointed out that in these estimates, PUREX has been under consideration; however, pyro-processing is more costly than PUREX and details can be referred in Ref. [6].

1.2 Methods of Reduction of the Radioactive Material

1.2.1 Disposals, Reposition and Transmutation

Reposition means a long-term disposal. Discussion about ways of reposition is of general interest, although a brief discussion related to economy of direct disposal has already been made in previous section. Very-long half-life and highly radiotoxic elements in the spent fuel are listed in Table 1.2 and among the fission products ^{99}Tc and ^{129}I are dominant in radio-toxicity and they exist even after thousands of years. Actinides ^{237}Np and ^{239}Pu need special attention to isolate them from interacting with the biosphere and proliferation. The issue has received high attention at both levels, i.e., academic and government administrations of several countries. Several ways of discrimination of LLNW have been proposed and those in use are presented in the following,

(i) **Ocean disposal**: It is continued from the year 1954 to 1993 by several countries like USSR, UK, Switzerland, USA, Belgium, France, Netherlands, Japan, Sweden, Russia, Germany, Italy, and South Korea. Presently, it is banned internationally. Regarding ocean disposal, a serious concern is reported about intensions of sinking of ships that carried radioactive waste [20].

(ii) **Direct injection**: It is continued by Russia and the USA and after the Blue Ribbon Commission, USA [21], only deep geological repositories are likely to be permitted. The basic requirement is to locate a geological stable formation and to excavate tunnels or a borehole mechanically 500–1,000 m below the bed where a vault or a strong room can be developed to isolate from human environment. Cases of illegal reposition are also recorded during 1980–1990 in African countries.

(iii) **Transmutation**: Large amount of plutonium produced from the uranium-fed reactors and the military projects is a matter of serious concern. In order to reduce its amount heavily, it can be used as a part of a reactor fuel in specially designed reactors or it can be incinerated or degraded to short half-life. As the cost of uranium is much smaller than any other alternative say Pu, makes Pu difficult to be used as a fuel. Another important concern is heavy accumulation of SNF, and inadequate number of available reprocessing plants makes the rate of reprocessing smaller than the requirement. It may be economical if SNF can be utilized in a reactor for energy production and to reduce the average half-life of SNF. The technology that can produce energy from the SNF, incinerate plutonium, or utilize it as a fuel needs to be developed, and it has been shown in several calculations as well as innovations that reutilization of SNF can be possible by the new technology of energy amplifiers in next few decades [22–27]. Energy amplifiers or accelerator-driven subcritical systems will be discussed in detail in later chapters.

(iv) **Space disposal**: Disposing off SNF into space may save planet earth although it is a costly affair. Lagrangian points have been discussed as safer place of disposal. In this process, many flights will be required for the disposal of entire SNF. In the space disposal, most serious problem may arise on failure of even a single flight and then secondary follow-up problems will further complicate it due to non-availability of single international regulation in the event of spill of the waste in several countries.

1.2.2 Other Applications

Radionuclides, both actinides and FP, have found several applications in medical science, radiology, industry, and education. Some of the application areas have reached to a status of a professional subject. Being within the scope of the book, some of the applications are summarized as follows.

1.2.2.1 Nuclear Medicines and Diagnostic Applications

It is a branch of medical imaging where a small amount of radionuclides are used to diagnose and or to treat several diseases. Many of the radioisotopes received from reprocessing of a SNF and some produced by the accelerator beams directly are used in several medical treatments, curing, relief in pain and diagnosing a disease like cancer or to get status in curing a disease. All such applications are identified under the title of nuclear medicines. Nearly 80% of such isotopes are extracted from the SNF during separation of FP, and 19% are produced using beams from cyclotrons and linear accelerators. Remaining 1% is produced by activation reactions. ^{99m}Tc is highly used in diagnosis and obtained from fission product, ^{99}Mo [28]. One is advised to refer [29, 30] for more details of extraction of ^{99m}Tc. In Table 1.5, some of the details of a few nuclear medicines are given along with their source of production.

1.2.2.2 Industrial Applications

Neutron-, gamma-, X-ray-, beta-, and alpha-emitting radioisotopes are frequently used in industries for sterilization, germination, monitoring defects, voids and moisture content, explorations of ores and minerals and calibration, etc. Portable neutron sources such as Am + Be, Po + Be, and Pu + Be are produced where the α + Be nuclear reactions become the cause of neutron generation and the thermalized neutron sources are generated on scattering with hydrogenous compounds. Such sources are used in investigations related to geological survey of minerals, oil and petroleum for petrochemical industry. Radioactive materials are being used also in security searches of precious metals and toxic nitrogenous materials on ports and other places of high security. Also, processes of transmission of gamma rays are used to determine the elemental content of ash of coal by measuring absorption coefficient, μ on line of coal carried on a conveyor belt. It may be mentioned that the ash has higher concentration of elements of higher atomic number than the natural coal. Mineral concentration is also determined by gamma scattering. Similarly, for measurement of plastic film thickness, beta-emitting isotopes are used.

In modern scientific industrial researches, polymerization has made a very important place. Polymer industry has even replaced successfully metal applications in a large number of products like guns. Hard and lightweight polymer sheets are developed from catalytic action of radiation,

Table 1.5 Some of the commonly used nuclear medicine isotopes, their special usage, and production resource [31]

S. No.	Isotope	Usage	Resource	S. No.	Isotope	Usage	Resource
1	^{11}C, ^{13}N, ^{15}O, ^{18}F	P.E.T; localizing epileptic focus, dementia, psychiatry, and neuropharmacology	Accelerator	11	^{131}I	Treating thyroid cancer, diagnosis of abnormal liver function, renal blood flow	Reactor
2	^{18}F (FDG)	Detection, monitoring progress of cancers during treatment	Accelerator	12	^{201}Tl	Diagnosis of heart muscle death and low-grade lymphomas	Cyclotron
3	^{57}Co	Marker to estimate organ size and in vitro diagnostic kits	Cyclotron	13	^{165}Dy	Synovectomy treatment of arthritis	Reactor
4	^{64}Cu	Wilson's and Menke's diseases, PET imaging, and therapy of tumors	Cyclotron	14	^{169}Er	Relieving arthritis pain in synovial joints	Reactor
5	^{67}Cu	Therapy	Cyclotron	15	^{166}Ho	Treatment of liver tumors	–
6	^{67}Ga	Tumor imaging, localization of inflammatory lesions (infections).	Cyclotron	16	^{192}Ir	Wire form for use as internal radio-therapy source for cancer treatment	Reactor
7	^{68}Ga	PET and PET-CT imaging	Produced from ^{68}Ge	17	^{137}Cs	Sterilization of blood	Reactor
8	^{111}In	Brain studies, colon transit	Cyclotron	18	^{51}Cr	To label red blood cells and quantify gastrointestinal protein	Reactor
9	^{123}I	Diagnosis of thyroid function	Cyclotron	19	^{188}Re	β-irradiation of coronary arteries from an angioplasty balloon	Reactor
10	^{125}I	Cancer brachytherapy (prostate and brain)	Reactor	20	^{99m}Tc	Used for imaging of the skeleton and heart muscles in particular	^{99}Mo from Reactor

e.g., gamma and neutron. Thus, modifying materials to obtain required resistivity or conductivity, tensile, porosity, hardness, and even to magnetic behavior radioisotopes are in use.

Long life of a power supply or a battery has importance for a satellite and several other technologies. Tritium and nickel-63 can be used for beta-voltaic cells which will have low power but long life. Russia is implementing a project for the development of ^{63}Ni power sources [32, 33]. ^{238}Pu has high decay heat ~0.56 W/gm, and it is being used as the radioisotope thermoelectric generators (RTGs) for the last several decades in USA. The heat is produced by stopping of intense alpha and the gamma rays in small thickness layer [32].

1.2.2.3 Laboratory Sources for Education

For laboratory education and other research problems related to nuclear science and technology, several radioactive sources retrieved from the nuclear waste of a reactor are used world over. From reactors following gamma, beta, and alpha sources are synthesized [33, 34]. In gamma, beta, and alpha spectrometry which are part of the nuclear industry following sources are used. They have small activity ~μCi.

(i) Gamma radionuclides: ^{60}Co, ^{133}Ba, $^{152,154,155}Eu$, ^{137}Cs, ^{134}Cs, ^{106}Ru
(ii) Beta emitters: ^{3}H, ^{14}C, ^{63}Ni, ^{55}Fe, ^{36}Cl, ^{41}Ca, ^{90}Sr, ^{99}Tc, ^{129}I, ^{135}Cs, ^{94}Nb

(iii) Alpha emitters (trans-uranic): $^{238-241}Pu$, ^{241}Am, $^{234,244}Cm$, ^{237}Np

1.3 Issues Related to Nuclear Waste

1.3.1 Proliferation and Security

The objective of IAEA is to 'accelerate and enlarge the contribution of atomic energy to peace, health and prosperity throughout the world' and for this purpose, it may regularly check inventories, take samples and analyze materials related to civil nuclear facilities. IAEA provides safeguard regulations to deter diversion of nuclear material by early detection. The process is complemented by controlling the export of sensitive technologies from the countries like UK and USA by way of voluntary organizations such as 'Nuclear Supplier Group (NSG).' The NSG has 46 countries with it presently. The main focus concern of IAEA is not to allow enrichment of uranium beyond the necessity of commercial civil plants. Similarly, it controls production of plutonium by means of either the nuclear weapons or the civil nuclear energy programs. China, France, UK, Russia, and USA are the five countries declared having nuclear weapons. For the issue of proliferation and security following is frequently notified by the IAEA,

> Since the foundation of the United Nations in 1957, the International Atomic Energy Agency (IAEA) has promoted two, sometimes contradictory, missions; on the one hand, the agency seeks to promote and spread the use of civilian nuclear energy internationally; on the other hand, it seeks to prevent, or at least detect, the diversion of civilian nuclear energy to nuclear weapons, nuclear explosive devices or purposes unknown. The IAEA as operates a safeguard system as specified under Article III of the Nuclear Non-Proliferation Treaty (NPT) of 1968, which aims to ensure that civil stocks of uranium, plutonium, as well as facilities and technologies associated with these nuclear materials, are used only for peaceful purposes and do not contribute in any way for proliferation or nuclear weapon programs. It is often argued [35] that proliferation of nuclear weapons to many other states has been prevented by the extension of assurances and mutual defense treaties to these states by nuclear powers, but other factors, such as national prestige, or specific historical experiences, also play a part in hastening or stopping nuclear proliferation. Total 189 countries have already signed Non Proliferation Treaty (NPT) including the five nuclear weapon countries [35].

Besides, the four nations India, Pakistan, North Korea, and Israel have either acquired or presumed to have acquired nuclear weapons and as on today they have not signed the NPT. North Korea had signed NPT but withdrew in 2003 [35].

About the NPT, some of the nations who have acquired nuclear capabilities have put forward [35] the following views,

1. Gandhi Plan of 1988 regards NPT as inherently discriminatory in favor of the nuclear weapon states. It speaks about a timetable for complete nuclear weapons disarmament. It endorsed early proposals of a Comprehensive Test Ban Treaty and for an international convention to ban the production of highly enriched uranium and plutonium for the purpose of weapons. This is also known as the 'cut-off' convention.
2. For some years, especially under the Clinton administration of USA pursued a variety of initiatives to persuade India and Pakistan to abandon their nuclear weapon programs and to accept comprehensive international safeguards for all their nuclear activities. In this regard, the Clinton administration had proposed a conference of the five nuclear weapon states and Japan, Germany, India, and Pakistan.

   India refused this including similar previous proposals. Also, India countered it with a demand that other potential weapon states such as Iran and North Korea should be invited to the conference and that regional limitations would only be acceptable if they were accepted equally by China.
3. Another approach of 'capping' the production of fissile material for weapon purposes. This

will hopefully be followed by 'roll back.' India and the USA jointly sponsored a UN General Assembly resolution of 1993 which calls for the negotiations for a 'cut-off' convention. In case India and Pakistan join such a convention, then they would have to agree to halt the production of fissile materials for weapons and to accept international verification on their relevant nuclear facilities (enrichment and reprocessing plants). India became prepared to join negotiations regarding such a cut-off treaty under the UN Conference on Disarmament.

4. Looking at the limited scope of confidence building between India and Pakistan in the year 1990, each side ratified a treaty of not attacking another's nuclear installations and in 1991, they exchanged the list of locations of nuclear plants although the lists were not regarded fully accurate.
5. Looking at the security reasons, India dropped its support to CTBT in the year 1995 and in the year 1996, India attempted to block the Treaty itself. After its nuclear test in 1998, India proposed that its ratification may be conditional as the five weapon states have to agree for reduction in their nuclear arsenals. The UN Conference on Disarmament has also called upon both India and Pakistan immediately to accede to the Non-Proliferation Treaty, presumably as non-weapon states.
6. Iran who is a signatory of NPT in regards with civilian use of nuclear energy was not allowed to perform uranium enrichment in secret manner which is a violation of the safeguard obligations of United Nations Security Council.

India has also been discussed in the context of nuclear apartheid. India's stand is that nuclear issues were directly related to national security and nuclear weapons should be a necessary right for all nations so long as certain states are in possession of nuclear weapons.

As an honest approach following views needs to be considered for all future deliberations,

(i) No nation has right to threaten other in order to keep up sovereignty.
(ii) All nations have right to safeguard security of its people and boundaries with safety of earth being more important.
(iii) Civilian use of nuclear capability is permissible and there can be all checks and tests of enrichment programs of all nations equally.
(iv) SNF is a wealth pertaining to the earth and all efforts need to be made to reprocess for the energy programs only.
(v) 'Saving the earth from nuclear fear' program need to be initiated with de-activation of nuclear weapons.

Security issues

An independent group of experts on arms-control and non-proliferation was founded in January 2006 in the name of International Panel on Fissile Materials (IPFM) [36]. This comprises experts from 18 countries including both nuclear weapon and non-nuclear weapon states. Following panel members of the IPFM attended the meeting in the year 2013:

Harold Feiveson, Alexander Glaser, Zia Mian, Frank von Hippel (Princeton University, US), Pavel Podvig, Anatoli Diakov (Russia), M. V. Ramana, R. Rajaraman (India), Jean du Preez (South Africa), José Goldemberg (Brazil), Pervez Hoodbhoy, A. H. Nayyar (Pakistan), Rebecca Johnson, Patricia Lewis, Gordon McKerron (UK), Martin Kalinowski, Annette Schaper (Germany), Jungmin Kang (South Korea), Li Bin (China), Miguel Marín Bosch (Mexico), Arend J. Meerburg (Netherland), Paul Meyer (Canada), Seyed Hossein Mousavian (Iran), Ole Reistad (Norway), Henrik Salander, Johan Swahn (Sweden), Mycle Schneider (France), Masafumi Takubo, Fumihiko Yoshida (Japan) and produced the following findings,

1. In 2013, the global stockpile of nuclear weapons was estimated to be over 17,000 weapons. The USA and Russia together holding over 16,000 of these weapons, and

the other seven nuclear weapon states holding a combined total of about 1,000 weapons.

2. The global stockpile of highly enriched uranium (HEU) at the end of year 2012 is estimated to be about 1,380 ± 125 tons. This is sufficient for more than 55,000 simple, first-generation implosion fission weapons.
3. The global HEU stockpile has been reducing. Over the past two decades, about 630 tons of HEU has been blended down, mostly by Russia. The USA, which has eliminated about 141 tons of mostly non-weapon-grade HEU, has chosen to set aside 152 tons of excess weapons HEU for a naval fuel reserve. The USA, UK, Russia, France, and China have all stopped producing HEU for weapons as well as any other purpose, in some cases decades ago. The first four of these states have made official declarations to this effect, China has done so informally. In 2012, Russia announced that it was resuming limited production of HEU for naval and fast reactor fuel. India is also producing HEU for naval fuel. Pakistan is producing HEU for weapons. It is possible that North Korea also may be producing HEU for weapons.
4. The global civilian stockpile now exceeds the military stockpile. There are civilian plutonium separation (reprocessing) programs in the UK, Russia, Japan, India, France, and China. In July 2012, the UK announced plans to close its THORP reprocessing plant, at Sellafield by 2018. This would end reprocessing in the UK. The future of Japan's reprocessing program is unclear in the wake of the March 2011 disaster at the Fukushima nuclear plant.
5. Under the terms of the 2010 of the 'Action Plan on Nuclear Disarmament,' the NPT nuclear weapon states have agreed to cooperate on steps to increase transparency and develop verification capabilities related to nuclear disarmament and in particular to report information that can further the openness and verification. According to the action plan, the nuclear weapon states were expected to report to the NPT Preparatory Committee in 2014 on progress toward meeting these obligations.

The IPFM recommended that the nuclear weapon states could make baseline declarations of the total number of nuclear warheads in their possession as of a specific recent date with a commitment to subsequent annual updates well in advance to the next NPT Review Conference. Also, as part of their baseline declarations by the end of 2015, NPT nuclear weapon states could make the following as public.

- Total national holdings of plutonium and of HEU as of a specific recent date.
- Amounts of HEU and plutonium in other countries and any foreign-owned material in a country.
- The portions of their HEU and plutonium stockpiles available for IAEA safeguards.

Earlier in the 2010 final document of NPT Review Conference emphasizes for the nuclear weapon states 'to declare, as appropriate, to the International Atomic Energy Agency (IAEA) all fissile material designated by each of them as not required for military purposes and to place such material as soon as practicable under IAEA or other relevant international verification and arrangements for the disposition of such material for peaceful purposes.'

Finally, the NPT weapon states need to declare and place under IAEA safeguards the following,

- All plutonium and HEU in civilian use
- All plutonium and HEU recovered from excess weapons or its nuclear weapons' complex and declared excess for weapon purposes and
- All plutonium and HEU going to waste disposal sites.

1.3.2 Reutilization

Is there a possibility to utilize SNF as such in a conventional reactor or any other system and the radio-toxicity of the long life isotopes is reduced to very short life and the process stands economical? Certainly, the question indicates toward the high amount of the leftover fuel elements to be burnt and LLA and LLFP to be incinerated drastically. Obviously, the problem focuses on the issue of LLFP to be incinerated as well as to reduce or incinerate amount of the reproduced LLFP from the burning of fuel component of SNF. In principle, this will certainly need a system to have very high excess of neutrons beyond the energy economy of the system. In this regard, Jansen [25] reaches to a conclusion that in doing both, energy output can be maintained corresponding up to certain value of neutron multiplication factor, k and excess neutrons for transmutation can be possible in near to $k \sim 1$ system only.

It is known that the LWR produces nearly 95% of the uranium fuel, 1% actinides, and remaining 4% FP. According to Lerner [37], alternatively, using the pyro-processing route of reprocessing all the trans-uranic elements of the spent fuel can be separated from FP for reutilization in a fast reactor without loading of FP of the SNF in the fast reactor. To expedite reutilization, a pyro-processing plant can be installed in between a LWR and the fast reactor. This can be promoted because pyro-processing-based technology [38] has following advantages,

(i) Several times more utilization of nuclear fuel than just 5%.
(ii) Uranium supply to continue to the second down the line reactor.
(iii) Minimization of risk of proliferation because of the integrity of uranium and higher actinide fuel elements.
(iv) This saves time and cost of transport and reposition. Also, risk period is reduced from thousands of years to hundreds of years.

Another approach being worked out for the last 15–16 years is IV generation reactors [39] with a feature that fuel will be recycled without separation of ^{239}Pu. Similarly, there are proposals of molten salt reactors (MSR) for utilization of thorium.

1.3.3 Radiation Damage and Gas Production

Reposition or even storage of SNF not only spreads radioactivity but produces gases as the decay products along with that are produced by activation or other nuclear reactions. For example fission products may emerge as the gases directly. Neutrons may also activate certain constituent like Ni and Al to produce alpha particles, which on emergence may accept electrons from the surrounding to act as helium gas. Thus, Ni of the steel canisters is reduced to iron and corrosion is promoted. Through the following corrosion reactions on one hand, depletion of nickel of steel takes place and this reduces the strength of the steel and on the other hand, iron gets oxidized. This leads to corrosion and production of hydrogen and helium gases.

$$\left.\begin{array}{l} {}^{58}_{28}\mathrm{Ni} + {}^{1}_{0}n \rightarrow {}^{59}_{28}\mathrm{Ni} \\ {}^{59}_{28}\mathrm{Ni} + {}^{1}_{0}n \rightarrow {}^{4}_{2}\mathrm{He} + {}^{56}_{26}\mathrm{Fe} \end{array}\right\} (E < 0.1\,\mathrm{MeV})$$

$${}^{58}_{28}\mathrm{Ni} + {}^{1}_{0}n \rightarrow {}^{4}_{2}\mathrm{He} + {}^{55}_{26}\mathrm{Fe}\ (E > 0.1\,\mathrm{MeV})$$

$${}^{59}_{28}\mathrm{Ni} + {}^{1}_{0}n \rightarrow {}^{1}_{1}\mathrm{H} + {}^{59}_{27}\mathrm{Co}\ (E < 0.1\,\mathrm{MeV})$$

Production of helium leads to grain boundary embrittlement and enhancement of swelling rate due to bubble formation. Thus, the high nickel steels become brittle for low doses too. In fact, trapping of rare gas atoms on the vacancy defects retards the annealing [40] and radiation defects become stable. This may induce brittleness. Issue of radiation damage and radiation resistant materials will be discussed in a dedicated Chap. 7 later in the book.

1.3.4 Protection of Society and Environment

The nuclear fuel before loading and after discharge from a reactor influences the environment. Also, small amount of radioactivity goes out of the reactor along with the rejected water and the steam that leak out of the reactor and ultimately it is absorbed by the soil or the crop in vicinity. Right from the mining of yellow cake by in-situ leaching process and the uranium ore grinding processes which is followed by the leaching the leftover tailings either to go to the soil or in water. In the whole process of leaching, thousands of m^3 of water is used and after extraction, water carries small trace of radioactivity which ultimately goes to far distant places during rains and in the dwellings. Effect of accidental release of excess radiation beyond dose limit in case of accident has shown enhanced cases of cancer patients in the surrounding of the plants.

According to IAEA safety standards [41],

> Regulating safety is a national responsibility. However, radiation risks may transcend national borders, and international cooperation serves to promote and enhance safety globally by exchanging experience and by improving capabilities to control hazards, to prevent accidents, to respond to emergencies and to mitigate any harmful consequences. States have an obligation of diligence and duty of care, and are expected to fulfill their national and international undertakings and obligations.

The IAEA declared the following 4 Safety Standard Committees (SSC) for the preparation and review of safety standards,

i. for nuclear safety (NUSSC)
ii. radiation safety (RASSC)
iii. the safety of radioactive waste (WASSC), and
iv. the safe transport of radioactive material (TRANSSC).

The Commission on Safety Standards (CSS) oversees the programs of IAEA safety standards. For safety and protection from radiation, ten principles are envisaged. Under its 7th principle highlighted as 'Protection of present and future generations' means that people and the environment, present and future, must be protected against radiation risks. In the following various important safety majors that are adopted in different nations having nuclear energy have been summarized.

United States of America

Nuclear Regulatory Commission (NRC) [42] of USA has ensured safeguards and security by regulating licensees by way of

(a) accounting systems for special nuclear and source materials, and
(b) security programs and contingency plans.

Their responsibilities include the following,

- Domestic safeguards
- Information security
- Radioactive material security.

Office of Nuclear Material Safety and Safeguards (NMSS) develops and implements NRC policy for the regulation and safe management and disposal of spent fuel and HLW. In the USA, Uranium Mill Tailings Radiation Control Act (UMTRCA) is also invoked in the year 1978 as the US Environmental Law. This authorizes the Environmental Protection Agency authority to establish health and environmental standards for the stabilization, restoration and disposal of uranium mill waste.

India

Atomic Energy Act, 1962, controls all nuclear activities in the country. In 1983, Atomic Energy Regulatory Board (AERB) came in existence to carryout regulatory and safety activities under the act. The regulatory authority of AERB is derived from the rules and notifications promulgated under the Atomic Energy Act, 1962. Environmental (Protection) Act, 1986, is added later on.

Also, Government of India invoked 'Atomic Energy (Radiation Protection) Rules, 2004' replacing the Radiation Protection Rules 1971. For more details of safety disposal of nuclear waste (SDNW), weapons of mass destruction (WMD), radiological dispersal devices (RDD), Foreign Trade Development and Regulation Act, 1992 (FDRA) and Nuclear Controls & Planning Wing (NC&PW) are existing. For several other outfits, reader is advised to refer to Nuclear Security in India [43].

France

In France, Transparency and Security in the Nuclear Field Act No. 2006-686 was invoked in June 2006 [44] under the Nuclear Safety Authority which is an independent administrative authority. In signing the act Ministry of Ecology is involved. The authority participates in the surveillance of nuclear safety and radiation protection and for informing the public in these fields. Also, separately, in 2006 Program Act on the Sustainable Management of Radioactive Materials and Wastes was passed to focus largely on environment in connection with the waste management [45].

United Kingdom

In UK, Office for Nuclear Regulation (ONR) is in charge of regulating nuclear sites but the legal responsibility of ensuing nuclear safety lies on the licensee. Government of UK is responsible for nuclear policy through its legislative regulatory framework. ONR sets the regulatory standards. According to statement of ONR Chief Executive Officer [46], 'ONR is not responsible for delivering a safe and secure nuclear industry; this is the responsibility of the nuclear industry itself.'

Also, Environment Agency [47] is an executive non-department public body which is sponsored collectively by the departments of environment, food, rural affairs, and it works to create better places for people and wildlife and support sustainable development of nuclear sites. Similarly, there is Scottish Environment Protection Agency.

China

In China, National Nuclear Safety Administration (NNSA) is a central government agency responsible for regulating nuclear safety, supervision on all civilian nuclear infrastructures in China and has authority of inspecting nuclear safety activities and to regulate the approval mechanism. It was established in the year 1984 under the State Science and Technology Commission, and since then, it has seen several controls. In the year 1998, it was transferred to State Environmental Protection Administration (SEPA). In the year 2008, SEPA was upgraded to the full-fledged Ministry of Environmental Protection and the NNSA has been working under the ministry. The Ministry is the nation's environmental protection department charged with the task of protecting China's air, water, and land from pollution and contamination [48]. It also serves as China's nuclear safety agency.

Russia

Nuclear Safety Institute (IBRAE) under the Russian Academy of Sciences [49] was established in the year 1988 with the aim of advancing basic research to address the problems of increasing safety of nuclear power plants. It is responsible for the (i) safety of nuclear power and industry, (ii) emergency response and radiation monitoring, (iii) strategic planning of back end of nuclear and radiation hazardous objects, (iv) to provide technological platform for industry and energy integrity, safety and security. Federal Environmental, Industrial and Nuclear Supervision Service (Russian Federation) is also started.

Japan

In Japan under its Atomic Energy, Basic Law Atomic Energy Commission was established in the year 1956. After the Fukushima Nuclear Disaster in 2012, Nuclear Regulatory Authority (NRA) was established [50] replacing 'Nuclear and Industrial Safety Agency,' to provide Nuclear Regulation for People and the Environment. The NRA is an external organization of the Ministry of the Environment with a high degree of independence. After its constitution, NRA has been enhancing nuclear regulations in the following areas,

(i) Development of counter measures against severe accidents.
(ii) Introduction of back-fit systems that all nuclear reactor facilities shall meet all new regulatory requirements.
(iii) Introduction of a 40-year operational time limit for nuclear reactor facilities from the time of start of a facility.

Pakistan

Pakistan has established Pakistan's Nuclear Security Regime, and it covers nuclear materials including radioactive materials, several associated activities, and facilities for their life cycle through following three pillars (i) legislative and regulatory framework (ii) institutions and organizations and (iii) nuclear regulatory systems and measures [51]. Pakistan's Atomic Energy Commission and its Nuclear Regulatory Authority are under its Legislative and regularity framework. Strategic Plan's Division (SPD) works as secretariat of National Command Authority (NCA) and develops technical solutions, Personnel Reliability Programme (PRP), and elaborate intelligence and security setups to deal with issues related to nuclear security, non-proliferation and Weapons of Mass Destruction (WMD) terrorism.

References

1. Nifenecker, H., Meplan, O., David, S.: Accelerator Driven Subcritical Reactors. IOP Publishing Ltd. (2003)
2. Yasin, Z., Iqbal, J., Shahzad, M.I.: Comparison of radionuclide's inventories and activities with slightly enriched uranium and plutonium fuel in CANDU reactors. J. Nucl. Sci. Technol. **1**, 31–36 (2011)
3. Raghupati, S., et al.: http://www.efn-uk.org/nuclear/nuc-lib/reactor-reports/index_files/India-PFBR-design.pdf and Puthiyavinayagam, P.: 48th Annual Meeting of TWGFR, IAEA IPPE, Obninsk, Russian Federation. https://www.iaea.org/NuclearPower/Downloadable/Meetings/2015/2015-05-25-05-29NPTDS/Country/11_TWGFR2015_India_IGCAR.pdf. Accessed 25–29 May 2015
4. Arafat, Y., et al.: Radiotoxicity characterization of HLW from reprocessing of uranium-based and thorium-based fuel-11390. WM2011 Conference, Phoenix, AZ (2011). Accessed 27 February–3 March 2011
5. Clement, C.H.: Editor, authors on behalf of ICRP, Eckerman, K., Harrison, J., Menzel, H.-G., Clement, C.H.: Compendium of dose coefficients based on ICRP publication 60. ICRP publication 119. Ann. ICRP **41**(Suppl) (2012)
6. Abdallah, M., Chang, H.M., Hampton, D., Mahmood, A., Mahmood, F.: Calvin Wong: Energy and Energy Policy Closing the Nuclear Fuel Cycle: Reprocessing. http://franke.uchicago.edu/bigproblems/2012-berry-papers/team3.pdf. Accessed 2 Apr 2018
7. Phillips, C., et al.: Improving the estimates of wastes from the recycling of used nuclear fuel. WM2013 Conference, Phonix, Arizona, USA, 24–28 Feb 2013
8. Choi, E.-Y., Jeon, S.M.: Electrochemical processing of spent nuclear fuels: an overview of oxide reduction in pyroprocessing technology. Prog. Nat. Sci. Mater. Int. **25**(6), 572 (2015)
9. Spent Nuclear Fuel Reprocessing: NEA/NSC/WPFC/DOC 15. OECD (2012)
10. Mccombie, C.: International Perspective: Reprocessing, Storage, and Disposal. NAE National Meeting, Irvine (2003). Accessed 6 Feb 2003
11. http://www.world-nuclear.org/information-library/nuclear-fuel-cycle/fuel-recycling/processing-of-used-nuclear-fuel.aspx and Status and Trends in Spent Fuel Reprocessing. IAEA-TECDOC-1467. http://www.pub.iaea.org/MTCD/publications/PDF/te_1467_web.pdf. Accessed Sept 2005
12. Spent Fuel Reprocessing Options. IAEA-TECDOC-1587 (2008). http://www-pub.iaea.org/MTCD/publications/PDF/te_1587_web.pdf. Accessed Aug 2008
13. Banerji, S.: Physics of ADS for energy and transmutation. Inaugural Talk Delivered at Workshop on ADSS for Energy and Transmutation Held at Jaipur (2006). Accessed 23–25 Jan 2006
14. Dey, P.K., Bansal, N.K.: Spent fuel reprocessing: a vital link in Indian nuclear power program. Nucl. Eng. Des. **236**(7–8), 723 (2006)
15. Natarajan, R., Raj, B.: Fast Reactor Fuel Reprocessing—An Indian Perspective. In: Proceedings of GLOBAL 2005, Paper No. IL012, Tsukuba, Japan (2005). Accessed 9–13 Oct 2005
16. Shughart II, William F.: Why does not US Recycle Nuclear Fuel. http://www.forbes.com/sites/realspin/2014/10/01/why-doesnt-u-s-recycle-nuclear-fuel/#5b0e0e1c7db4. Accessed Oct 2014
17. Global Fissile Material Report 2013: Increasing Transparency of Nuclear Warhead and Fissile Material Stocks as a Step Toward Disarmament. Seventh Annual Report of the International Panel on Fissile Materials. www.fissilematerials.org
18. Orszag, P.R.: Costs of Reprocessing Versus Directly Disposing of Spent Nuclear Fuel. CBO Report (2007). https://www.cbo.gov/sites/default/files/cbofiles/ftpdocs/88xx/doc8808/11-14-nuclearfuel.pdf. Accessed Nov 2007 and UCS: Nuclear Reprocessing: Dangerous, Dirty, and Expensive. http://www.ucsusa.org/nuclear-power/nuclear-plant-security/nuclear-reprocessing#.V3X5ZNKgE0F
19. The Economics of Reprocessing vs. Direct Disposal of Spent Nuclear Fuel (Cambridge, Mass: Belfer Center for Science and International Affairs, John F. Kennedy School of Government, Harvard University, December 2003)
20. Mafia sank boat with radioactive waste: Official, AFP. Accessed 14 Sept 2009 and From cocaine to plutonium: Mafia clan accused of trafficking nuclear waste. The Guardian. Accessed 9 Oct 2007

21. Blue Ribbon Commission on America's Nuclear Future: Executive Summary. Accessed Jan 2012
22. Rubbia, C., et al.: Conceptual design of a fast neutron operated high power energy amplifier, CERN/AT/95-44 (ET) (1995). See also Rubbia, C.: A high gain energy amplifier operated with fast neutrons. AIP Conference Proceedings 346, International Conference on Accelerator-Driven Transmutation Technologies and Applications, Las Vegas (NV), US (1994)
23. Carminati, F. et al.: An Energy Amplifier for Cleaner and Inexhaustible Nuclear Energy Production Driven by a Particle Beam Accelerator. CERN/AT/93-47 (ET) (1993)
24. Bowman, C.D. et al.: Nuclear energy generation and waste transmutation using an accelerator-driven intense thermal neutron source. Nucl. Instr. Methods Phys. Res A320, 336 (1992)
25. Janssen, A.J.: Transmutation of Fission Products in Reactors and Accelerator-Driven Systems. ECN-R-94-001 (1994)
26. Salvatores, M., et al.: MUSE-1: A first experiment at MASURCA to validate the physics of sub-critical multiplying systems relevant to ADS. In: Proceeding of the Second International Conference on Accelerator Driven Transmutation Technologies and Applications, Kalmar, Sweden, ADTTA (1996). Accessed 3–7 June 1996
27. Kumar, V., et al.: Role of (n, xn) Reactions in ADS, IAEA-Benchmark and the Dubna CASCADE Code. Pramana-J. Phys, **68**(2), 315–324 (2007) (Also, the volume of Pramana-Journal of Physics is dedicated to studies related to ADSS)
28. Feasibility of Producing Molybdenum-99 on a Small Scale Using Fission of Low Enriched Uranium or Neutron Activation of Natural Molybdenum. Technical Reports Series No. 478. http://www-pub.iaea.org/MTCD/Publications/PDF/trs478web-32777845.pdf
29. OECD/NEA: A Supply and Demand Update of the Mo-99 Market (2012)
30. IAEA 2015: Feasibility of Producing Molybdenum-99 on a Small Scale Using Fission of Low Enriched Uranium or Neutron Activation of Natural Molybdenum. Technical Reports Series #478
31. Rao, D.V., Chandra, R., Graham, M.C. (eds.): Physics of Nuclear Medicines-Recent Advances. American Institute of Physics Inc., New York (1984)
32. Radioisotopes in Industry. http://www.world-nuclear.org/information-library/non-power-nuclear-applications/radioisotopes-research/radioisotopes-in-industry.aspx
33. Olsen, L.C., Cabauy, P., Elkind, B.J.: Betavoltaic power sources. Phys. Today (2012), 35–38 (2012). www.physicstoday.org
34. Hou, X.: Radiochemical Analysis of ^{3}H, ^{14}C, ^{55}Fe, ^{63}Ni in Waste and Environmental Samples. Risø National Laboratory for Sustainable Energy, Technical University of Denmark
35. Nuclear Proliferation. https://en.wikipedia.org/wiki/Nuclear_proliferation#Non-proliferation_efforts
36. International Panel on Fissile Materials: Global fissile material report 2013 increasing transparency of nuclear warhead and fissile material stocks as a step toward disarmament. www.fissilematerials.org
37. Lerner, L.: Nuclear fuel recycling could offer plentiful energy. http://www.anl.gov/articles/nuclear-fuel-recycling-could-offer-plentiful-energy
38. Williamson, M.: Pyro-processing technologies. http://www.cse.anl.gov/pdfs/pyroprocessing_brochure.pdf
39. Generation IV Nuclear Reactors. http://www.world-nuclear.org/information-library/nuclear-fuel-cycle/nuclear-power-reactors/generation-iv-nuclear-reactors.aspx
40. Matzke, H.J.: On the stabilization of radiation damage by the rare earth gases. J. Nucl. Mat. 30, 107 (1969)
41. IAEA Safety Standards Series No. GSR Part 3 (Interim). http://www.pub.iaea.org/MTCD/publications/PDF/p1531interim_web.pdf
42. United States Nuclear Regulatory Commission. http://www.nrc.gov/security.html
43. Nuclear Security in India. http://www.indianhighcommission.com.my/Pdf/NUCLEAR_SECURITY_IN_INDIA.pdf
44. Nuclear Safety Authority, Act 206, France. file:///C:/Users/lenovo/Downloads/loiTSN-uk.pdf
45. Program Act on the Sustainable Management of Radio-Active Materials and Wastes, France, 2006. file:///C:/Users/lenovo/Downloads/radioactive-waste-management-act-280606.pdf
46. Office for Nuclear Regulation. ONR. http://www.onr.org.uk/documents/a-guide-to-nuclear-regulation-in-the-uk.pdf
47. Environment Agency. https://www.gov.uk/government/organisations/environment-agency
48. Ministry of Environmental Protection of the People's Republic of China. https://en.wikipedia.org/wiki/Ministry_of_Environmental_Protection_of_the_People%27s_Republic_of_China
49. The Nuclear Safety Institute (IBRAE). http://en.ibrae.ac.ru/
50. Nuclear Regulation Authority. Japan. https://www.nsr.go.jp/english/e_nra/nsr_leaflet_English.pdf
51. Pakistan's Nuclear Security Regimes: http://www.mofa.gov.pk/documents/PNSR.pdf. Accessed 7 Apr 2018

2 Special Hybrid Systems and Molten-Salt Reactors

Complete utilization of heat and escape out neutron flux generated new concept of hybridization of two or more nuclear energy systems which is highly perspective in case of gen. IV reactors. It has high utility in fusion reactors where stopping of high-energy neutrons is almost impossible. Concept is being worked out at the international collaborations. Hybridization of nuclear and non-nuclear systems of producing electricity and utilization of heat resources are also discussed. Such concepts need lot of innovative procedures for integration.

2.1 Concept of Hybridization of Nuclear Energy

So far, two kinds of nuclear energy systems are known; one is for atomic fission energy, and other one is for nuclear fusion energy. An atomic energy reactor, having ^{nat}U fuel a single fuel, and it is ignited by a small energy system, a neutron source. This energy input is negligible compared to fission energy. These reactors are also called as neutron poor and energy-rich systems. On the other hand, fusion reactors are neutron rich but energy poor systems. Energy released in one fission reaction is more than 10 times of one fusion reaction. Extra neutrons of a fusion reactor can better be utilized in another system based on fission process by providing blanket of a fissionable fuel outside the fusion reactor dimensions.

In principle in a fission reactor, there is no hybridization as there is an input of neutron energy and there is a single fuel system. Hybridization may be said to take place when there is a combination of fuels, e.g., uranium and plutonium mix or thorium and uranium mix fuels, or there is a mechanism of generating a big neutron source such as spallation neutron source for running a fission reactor for both energy and transmutation purposes. In the latter case, a high-energy spallation neutron generates a large number of neutrons which produce fission reactions as well as transmute the specially organized HLW also.

Another possibility is to combine fusion and fission reactors for utilization of excess neutrons of fusion reactor to enhance energy production by providing a fission material blanket around the fusion reactor [1]. The blanket can have uranium or thorium as a fuel. There can be several other ways of enhancing overall efficiency of the energy system or making a cluster of utilization of any extra availability of resource economically.

2.1.1 Hybrid Nuclear Energy Systems

Fusion reactors are operated on two basic mechanisms. One related to magnetic confinement and other on laser. In case of magnetic confinement of plasma, say, deuterium and tritium plasma, the process of fusion is accelerated

V. K. Verma and K. Katovsky, *Spent Nuclear Fuel and Accelerator-Driven Subcritical Systems*, Green Energy and Technology, https://doi.org/10.1007/978-981-10-7503-2_2

by introduction of accelerated deuterium ion of energy, $E_h \sim 100$ keV. The deuterium ion energy is much above ~ 10 keV temperature of the desired plasma. To get the accelerated ions, one needs small energy accelerators to produce deuterium ion. Efficiency, ν_d, of the accelerators, i.e., ratio of KE of accelerated ion, E_h, to the electric energy supplied to the accelerator is $\sim$ 60–70%.

A figure of merit of the fusion energy system can be defined as the ratio of fusion energy, E_{Fu}, to the given heat energy, E_h,

$$Q = E_{Fu}/E_h \tag{2.1}$$

Here, E_h is required to heat the plasma. Initially, E_{Fu} is in the form of KE of produced alpha and neutron in a (*d*, *t*) plasma system, and it is required to be converted to the heat energy. In case of toroidal Tokamak, large value of Q is expected to be available.

More of the E_{Fu} energy is converted to the heat; factor Q will be higher. Thus, there is a need of a mechanism to be developed for the utilization of kinetic energy of charged particle introduced in the plasma to initiate fusion reaction. In turn, it produces heat on one hand and number of neutrons on the other hand which will be multiplied further by way of a fission reaction in a fusion–fission hybrid system.

Initially, for an elementary process, Q is expected to be $\ll 1$ so long as reaction energy in the form of kinetic energy of product nuclei is not utilized in internal collisions.

$$\text{Thus, heat of the system at a time} = E_h + E_{Fu} = (1+Q)E_h \tag{2.2}$$

The fusion energy, E_{Fu}, ultimately has to be converted into the electric energy. If the conversion efficiency is ν_e, then the electric energy output can be written as,

$$Q' = (1+Q)\nu_d \times \nu_e \sim \frac{1}{4}(1+Q) \tag{2.3}$$

A fusion energy system will be called efficient if $Q' \geq 1$. This is possible when $Q \geq 3$. This can be achieved by thermalizing the available kinetic energies of the nuclei produced in the fusion reaction (see Eq. 2.4). Among the existing fusion energy sources working with the magnetic confinement are called as Tokamak [2]. One Tokamak is shown to have plasma performance with $Q = 1$.

Deuterium and tritium are the main ingredients of most of the fusion reactors because of availability of the two reactants. Deuterium is a stable isotope of hydrogen found in water. Tritium is produced in a reactor. Also, it can be produced by reaction of a neutron with Li^6. Isotope Li^6 is abundantly available in mineral ores. At the same time, (*d*, *t*) fusion reaction is favored from the point of lower temperature requirement as can be deduced from the data shown in Fig. 2.1. As a matter of mechanism, when the two nuclei deuterium and tritium are brought close enough to fuse it gives rise to production of an alpha and a high-energy neutron as shown in Eq. (2.4). Barrier penetration energy for the (*d*, *t*) reaction is ~ 280 keV [3], and similarly penetration energy is ~ 210 keV for (*d*, *d*) and ~ 580 keV for (*d*, ^{3}He) fusion reactions. The (*d*, *t*) fusion reaction is as follows:

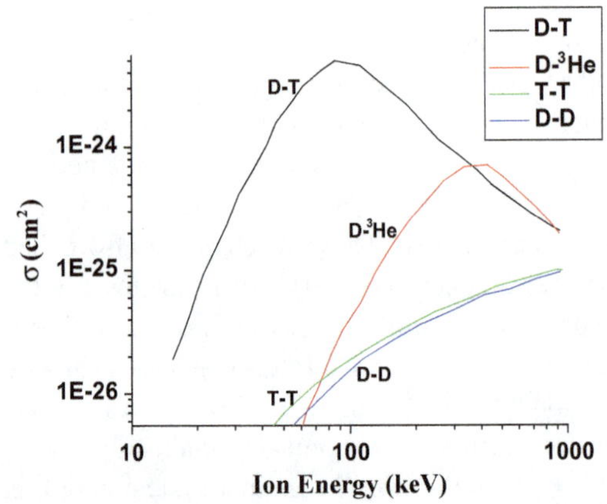

Fig. 2.1 Ion energy versus cross-sectional plot of different fusion reactions [2]

$$_1D^2 + {}_1H^3 \rightarrow {}_2He^4(3.52\ \text{MeV}) + n_0(14.06\ \text{MeV}) \quad (2.4)$$

According to Eq. (2.4), available total fusion energy, E_{FU} = 17.58 MeV, of which neutron energy share is ~80% and 20% is shared by alpha charged particle. The fusion energy E_{FU} needs to be converted into heat energy. The energy component of charged particle alpha is converted differently to the neutral neutron.

The reactions that may lead to the production of tritium are as follows:

$$^3Li^6 + {}_0n^1(\text{thermal}) = {}_2He^4(2.05\ \text{MeV}) + {}_1H^3(2.73\ \text{MeV}) \quad (2.5)$$

$$^3Li^7 + {}_0n^1(\text{fast}) = {}_2He^4 + {}_1H^3 + {}_0n^1 - 2.47\ \text{MeV} \quad (2.6)$$

There are several advantages of fusion reactors over the fission reactors, and they are taught in elementary classes, but, a fusion reaction is hard to be achieved in the first place because a small perturbation in reactor condition would probably terminate the reaction [4]. Thus, the basic challenges of a fusion reactor can be summarized as follows,

(i) Heating of a reacting mixture to a very high temperature is hard because energy goes in overcoming the repulsive force of positively charged nuclei. This is equivalent to a barrier penetration process. For example, as mentioned earlier also, this can be achieved by introducing deuterium ion of kinetic energy ~300 keV in case of (*d*, *t*) reaction. Introduction of high amount of such ions into the fusion system also enhances density of the plasma.
(ii) On compression of the mixture to a high-density, probability of collision, hence a reaction is enhanced.
(iii) To keep the reacting mixture together for long enough time for the fusion reaction and to produce energy at a rate that is greater than the rate of input energy provides both heat and compression.

Obviously, attaining a 'breakeven' is a technological challenge for sustaining the plasma condition of high temperature and density for longer time. The condition of sustainment can be derived from the 'Lawson criterion' that the product of number density of medium and confinement time $n \times \tau$ needs to be greater than 10^{14}. In the year 1991, European scientists have successfully sustained (*d*, *t*) plasma bursts for 2 s producing energy equivalent of 2 MW. Experimental Fusion reactor at Princeton sets a record of 1 s of 10.7 MW bursts in the year 1994. Following such successes, USA, Japan, Russia, and the European Union started designing and construction of the International Thermonuclear Experimental Reactor (ITER) facility of 500 MW to sustain for 400 s with $Q > 10$ or sustaining for 3000 s with $Q > 5$. Later on, China, India, and South Korea have also joined the program. Besides, working on hot plasma, several research groups are also working on 'cold fusion' of hydrogen at room temperature [4].

Nuclei produced as a result of several kinds of fusion reactions can be thermalized in dense medium, and as a result, very high temperatures are produced. Plasma in the given medium can be confined technologically by way of magnetic confinement or inertial confinement by lasers. Inertial confinement fusion (ICF) is a process to initiate nuclear fusion reactions by heating and compression of a fuel target which is typically kept in the form of a pellet containing a mixture of deuterium and tritium. In Fig. 2.2, a comparison of achievements of fusion energy over three decades by using either of the magnetic confinement or the ICF technology is shown in relation with progress of attaining computer power in the same time span.

In case of the ICF technique, impact of laser light starts compression by shock wave propagation which heats the outer layer of (*d*, *t*) fuel. Subsequently, compression of the inner target material [5, 6] starts the desired fusion reaction.

After the fusion energy is thermalized in a fusion system and plasma exceeds the loses, the condition, $Q > 1$ is achieved. Fusion reaction will sustain if fuel is continuously supplied. The plasma is confined by keeping the ions and the

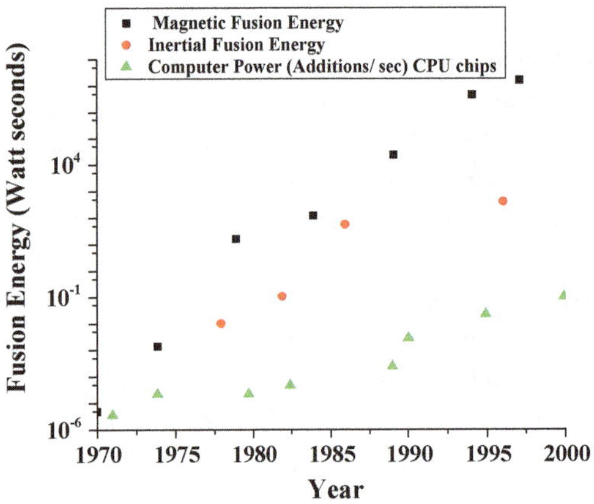

Fig. 2.2 A comparative study of development of magnetic and inertial confinement fusion [2]

electrons inside the system which is technically difficult.

2.1.2 Fusion–Fission Hybrids

A highly common design of fusion–fission hybrid system is shown in Fig. 2.3, where a fusion plasma core is surrounded by the fission blanket of U or Th [7, 8]. In the figure, a design proposal of ITER is projected for a 500 MW fusion power and neutron fluence $\sim 1.8 \times 10^{20}$ n/ s.

As mentioned earlier, a fusion–fission hybrid system necessarily has a fission fuel blanket outside the fusion system. A neutron being neutral has high chance to escape out from the fusion system, and it can be captured in the blanket material of high density of fission fuel and that leads to conversion of its energy into fission. In case of a 14.06 MeV, neutron escaping to the fission blanket may produce a fission and releases energy ~ 190 MeV/ fission. This energy is 13.5 times of energy of the neutron. In a (d, t) plasma system, energy, E_{Fu}, is shared with a fraction, $f_n \sim 80\%$ by neutron and remaining fraction, $(1 - f_n) \sim 20\%$ by a charged particle (cp), alpha in the final state of (d, t) reaction, for example. Thus,

$$\begin{aligned} E_n &= f_n \times E_{\text{Fu}} \quad \text{and} \\ E_{cp} &= (1 - f_n) \times E_{\text{Fu}} \end{aligned} \tag{2.7}$$

To realize the mechanism of total fusion energy utilization, it is necessary to understand the following schematic diagram as shown in Fig. 2.4 [9]. In the (d, t) fusion chamber, the alpha on colliding with the first wall of the fusion chamber, may be the vanadium metal, will transfer its KE to atoms of the wall causing both heat generation and radiation damage. Thus, the converted KE into the heat and the fission-heat of the blanket will be converted into steam; hence, the electricity or heat may directly be converted to the electricity by another mechanism. A part of the so produced electric energy, E_{rec}, may be fed back to the ion accelerator. Efficiency of the accelerator, η_1, for converting electric energy,

Fig. 2.3 A most commonly discussed fusion–fission hybrid system under the magnetic confinement [2, 7]

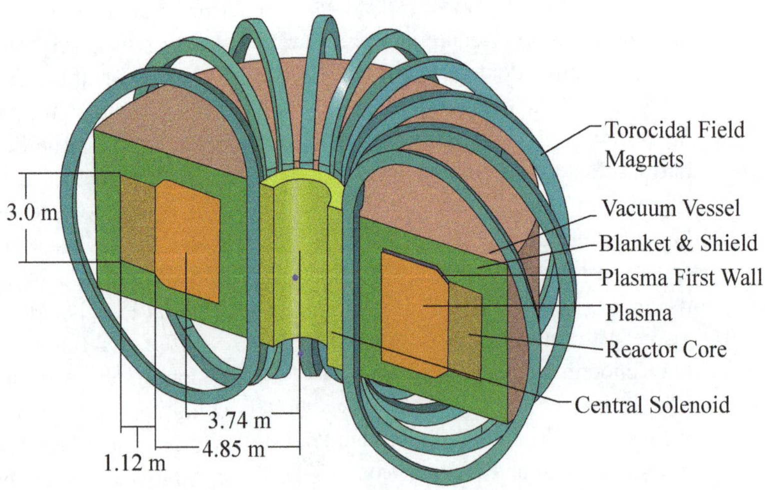

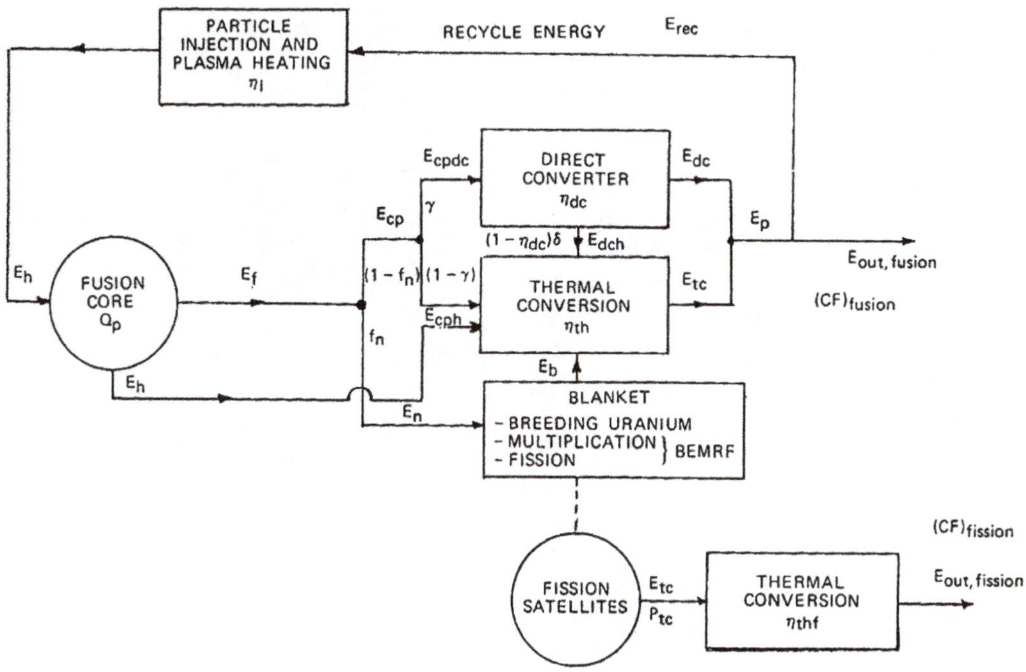

Fig. 2.4 A schematic diagram of fusion energy conversion into electric energy and its internal utilization [9]

E_{rec}, into KE of an ion, E_h, can be written as follows:

$$\eta_l = \frac{E_h}{E_{rec}} \tag{2.8}$$

Similarly, fraction of fusion energy, $(1 - f_n)$, carried by the charged particle like alpha will be converted into heat by a factor, γ; thus,

$$E_{cpdc} = \gamma E_{cp} \tag{2.9}$$

Further, the E_{cpdc}, heat energy, can be converted directly to electricity by say thermoelectric generator, or it can be converted first to steam and then to electricity by using a turbine as mentioned earlier. The two routes of conversion into electricity are shown as outgoing channels of E_{out} through 'direct converter' and the 'thermal conversion.' In future works, advance systems of such conversions may be investigated to have better efficiency.

Neutron energy, E_n, is further multiplied by way of the fission energy of fuel through ^{233}U. A part of E_n is also utilized in neutron multiplication. If BEMRF is the 'blanket energy multiplication factor,' i.e., ratio of energy multiplied in blanket to the source neutron energy [9], then the energy in the blanket system may be written as follows:

$$\begin{aligned} E_b &= \text{BEMRF} \times E_n \\ &= \text{BEMRF} \times f_n \times E_{Fu} \end{aligned} \tag{2.10}$$

Besides, as mentioned earlier, there is a chance that neutrons are multiplied in the blanket by way of non-fission reactions or energy deposited in the blanket by other non-fission reactions. In turn, they produce extra energy on interaction with the fission fuel material. This leads to

$$\text{BEMRF} = \text{BEMR} + \frac{\text{no. Th}(n,f) \times 184.2}{\text{Source } n - \text{energy (MeV)}} \tag{2.11}$$

For a detailed discussion of contributions of different conversion systems and parameters of the hybrid system, reader is advised to refer to Ragheb and Singh [9].

2.2 Reactor with Liquid Fuel

^{232}Th isotope is commonly known as 'thor,' and it is found as a 100% isotopic composition in the form of thorite ($ThSiO_4$) and thorianite (ThO_2+UO_2). It is 3 times more abundant than uranium in nature. Thorium is also found in ash of coal. Thorium fuel-based fusion–fission hybrid reactor is expected to be a sustainable source of energy for future. The use of the molten-salt reactor (MSR) was pioneered at Oak Ridge National Laboratory first time. At ORNL, two prototypes of molten-salt reactors, namely Aircraft Reactor Experiment (ARE) and Molten-Salt Reactor Experiment (MSRE), were successfully designed, constructed, and operated from the year 1965–69. Later on, the two experimental reactors were suspended from operation for a long period of time in USA.

Two possibilities of using molten salts, one comprising Na–Th–F–Be (as NaF.BeF_2.ThF_4) with the (*d*, *d*) neutron source and second Li–Th–F–Be (as LiF.BeF_2.ThF_4) salt with (*d*, *t*) source, have been studied. In such hybrid system having thorium as the target blanket and graphite as the neutron reflector, the (*d*, *t*) reaction provides 0.74Th (*n*, γ) breeding to ^{233}U compared to 0.76Th (*n*, γ) by the (*d*, *d*) neutron source [10, 11]. As shown with the help of Eq. (2.6), tritium is also produced in neutron interaction with ^{6}Li of a molten salt. Subsequently, this affects breeding of ^{233}U in a (*d*, *t*)—fusion reactor because of further generation of a part of neutron flux. The heat advantages in the two cases of plasmas are comparable and enormous. Such reactors can also be named as Molten-Salt Breeder Reactor (MSBR) [12]. Shortly, a MSR has advantages mainly of safety, economy, and efficiency and disadvantages of design difficulties which need separate discussions with details.

2.2.1 Single-Fluid Reactor

Single-fluid prototype reactor was designed at ORNL [13] wherein fluoride salt containing thorium and uranium was filled in a large reactor vessel with graphite rods for the purpose of moderation as shown in Fig. 2.5. In the ORNL MSBR reactor, graphite is reduced near the outer wall of the vessel hence allowing better chance of capture of thermal neutrons by the thorium fuel in the area. This also allowed lesser neutron leakage. Being a liquid fuel, one such reactor has also provisions of removal of fission products and that reduces cost of reprocessing. This will, however, require periodic filling of uranium fuel.

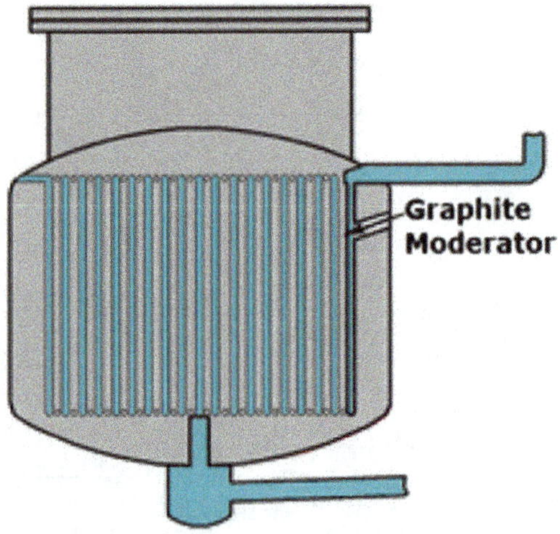

Fig. 2.5 A schematic diagram of single-fluid reactor as one of the liquid fluoride thorium reactors (LFTRs) [13]

2.2.2 Double-Fluid Reactor

A double-fluid reactor employs two separate liquid fuels. One is the core fuel, and another fuel is in the blanket of the core. The core contains high neutron density to burn ^{233}U. The thorium fuel salt in the blanket absorbs even fast neutrons to convert into ^{233}Pa which slowly decays to ^{233}U. There is lesser chance of ^{233}U to capture slow neutrons to convert into ^{234}U because of the lower amount of slow neutron flux. ^{233}U is recovered by way of fluorination process then allowed for the vacuum distillation, and in this process, fission products are left in the bottom of the blanket as the waste. The ^{233}U fuel so recovered can be filled in the core for burning as a fissile fuel. Recently, Holcomb [14] has discussed various MSR variants with different fuels, coolant possibilities, challenges and advantages.

The design has advantage in the simplified fuel processing because of low fissile inventory being located in a small core. Also, it provides more efficient breeding opportunity as the Th-blanket captures neutrons leaked out from the core. This results in more breeding from a small reactor. The design, however, has weakness in the form of barrier wall between the two fluids as it needs frequent replacement because of its fast radiation damage.

Further, Shimazu [15] has discussed three core concept of FUJI-U3 design having fuel salt $LiF–BeF_2–ThF_4–^{233}UF_4$ where the fuel to graphite ratio is kept as 0.39, 0.27, and 0.45 in core 1, core 2 and core 3 respectively, for a 200 MWe reactor. The reactor is to be driven by 1 GeV and 200–300 mA proton accelerator. The reactor system will have possibility of multi-beam funnelling to achieve the designed beam current. Life of the FUJI-U3 system with graphite moderator is estimated to be 20–30 years.

2.3 Other Hybrid Energy Systems

Concept of hybridization when extended to other energy systems then it opened up several different possible options of integration of energy inputs like nuclear and other renewable systems or the energy outputs like electricity, gasoline, water and heat etc. using proper complimentary energy conversing processes [16]. For example, different input energy systems like nuclear and solar energy systems may feed to a master grid system after having required input parameter of frequency. By enabling more than one energy conversion units, Nuclear Hybrid Energy Systems (NHES) may provide additional opportunities of flexible energy management, delivering various types of ancillary services such as operating reserves, regulating, ramping, load following and supplemental reserves and for enabling operational flexibility for value, technical, and/or economic optimization. Chen et al. [17] have considered the issues of variability in depth in accordance with the two models that are summarized in the following.

2.3.1 West Texas Model of NHES

West Texas model assumes inputs like nuclear, wind, natural gas, and water resources and to have outputs as electricity and gasoline, etc. In practice, it combines a small nuclear system with wind turbine park to produce electricity and to convert carbon resources to the gasoline using excess thermal capacity as steam. This has a flexible generation resource and a flexible thermal load. From the West Texas model, shown in Fig. 2.6, it can be understood that not only the electricity from the small modular nuclear reactor (SMR, 180 MW) plant and the wind energy turbine (45 MW) electricity can be sent to the electric grid after proper matching of parameters but extra steam of the nuclear plant (conversion capacity $\sim$45 MW) can be utilized to generate LPG through a gasoline production plant (GPP). SMR and wind power have maximum capacities as 180 and 45 MW, respectively, and the heat deducted in the form of steam for utilization for the LPG production is also variable with maximum demand of 45 MW. Ultimately, supply $\sim$180 MW can be sent to the grid with a little management of wind power and GPP.

2.3.2 Northeastern Arizona Model of NHES

Northeastern Arizona model assumes inputs like nuclear and solar photovoltaic (PV) systems to supply electricity to the grid, and at the same time, it produces fresh drinkable water from the saline water. The Arizona NHES has a flexible electrical load and can be operated as a flexible generation resource and flexible load resource. In Fig. 2.7, schematic diagram of the Arizona NHES model is shown. In the figure, power sharing is shown, and it may be seen that ultimately a lesser power $\sim$135–165 MW electricity is fed to the grid compared to the West Texas Model of NHES. The loss of electricity to the grid has been evaluated considering total economics with respect to the region.

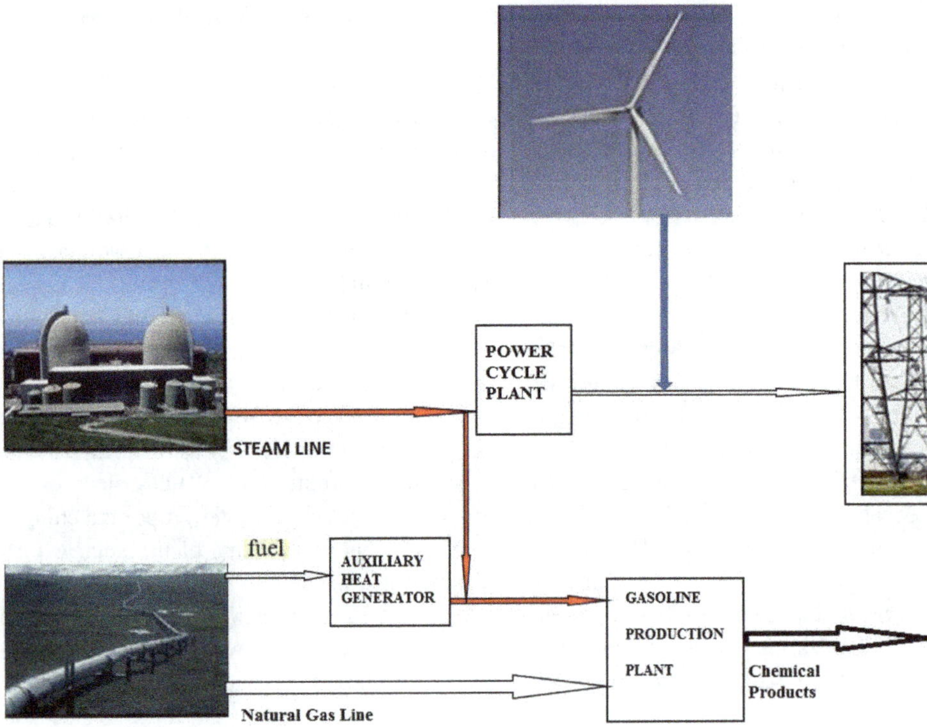

Fig. 2.6 A schematic diagram of West Texas model of NHES

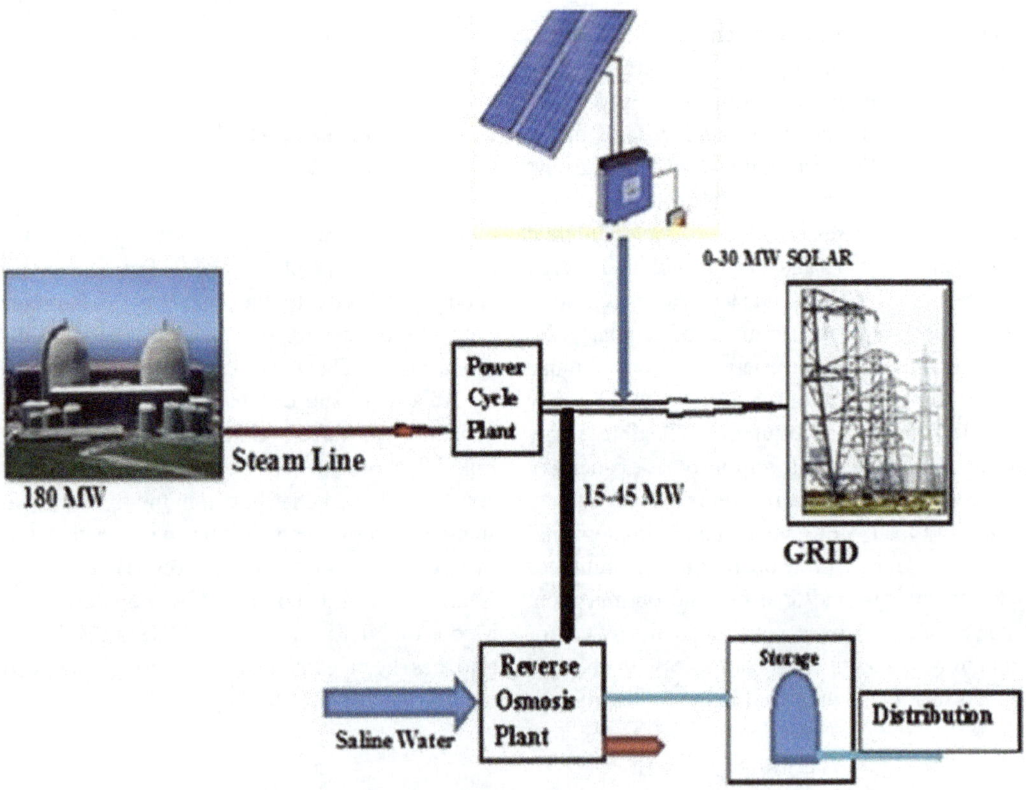

Fig. 2.7 A schematic diagram of the Arizona NHES model

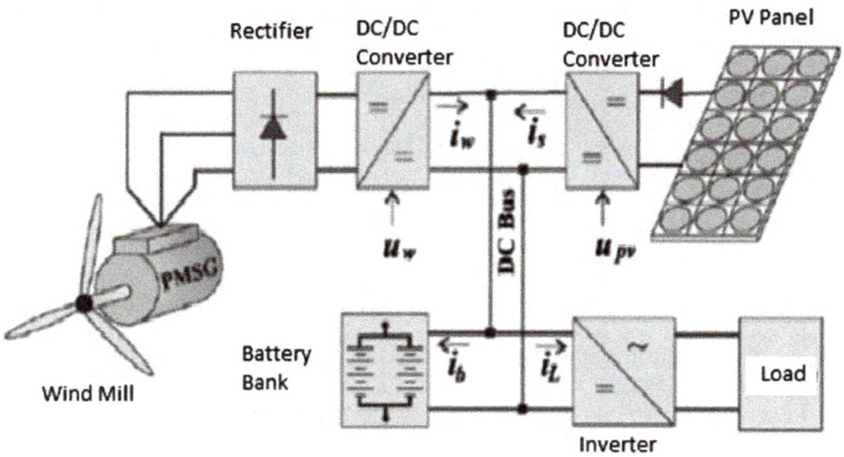

Fig. 2.8 A wind mill and photovoltaic hybrid energy system

Economic functions and operation optimization conditions are worked out and discussed in detail by Chen et al. [17].

2.3.3 Wind and PV Hybrid Energy System

Valenciaga and Puleston [18] have explored possibility of hybrid system of wind and photovoltaic (PV) system and a stand-alone supervision control system. A 'wind energy conversion systems (WECS)' comprises a windmill, multipolar synchronous generator (PMSG), a rectifier, and a dc/dc converter for interfacing. PMSG and control is discussed in details by Melicio et al. [19]. The solar module comprises of several solar panels connected to the dc bus as shown in Fig. 2.8. The battery bank and its cost play a fundamental role in the overall system cost [20].

Apparently with the growth of hybrid energy systems and efficient utilization of resources, energy parks will be appearing on the world map. This may also initiate the process of clustering of industries related to energy resources. This may, however, present a different scenario where nuclear energy and other energy resources as well as resource utilization will stand erect in the same area. This may raise several issues related to security, proliferation, and activating new regulations.

References

1. Bethe, H.A.: The fusion hybrid. Phys. Today **32**(4), 44 (1979)
2. A Report of the Conference on Hybrid Fusion-Fission Systems. Washington DC (2009). web.mit.edu/fusionissi on/HybridsPubli/Hybrid_Fusion_Fission_Conference_A.pdf. Accessed 19–20 May 2009
3. Bosch, H.-S., Hale, G.M.: Improved Formulas for Fusion Cross-Sections and Thermal Reactivities. http://iopscience.iop.org/0029-5515/32/4/I07 and http://kfe.fjfi.cvut.cz/~valenpe7/files/12FVHE/BoschHale.pdf
4. Fusion Energy.: National Energy Strategy, Executive Summary, p. 258. https://www.ems.psu.edu/~radovic/Chapter14.pdf. Accessed 1991/1992
5. Nakatsutsumi, M., et al.: Investigations of High-Energy-Density State at European XFEL. ICHED 2015 abstract book (2015), San Diego, California. https://www.researchgate.net/profile/Gerd_Priebe/publication/302596511_ICHED_2015/links/57c5824108ae7642019b0709.pdf?origin=publication_list. Accessed 23–27 Aug 2015
6. Kodama, R., et al.: Fast heating of ultrahigh-density plasma as a step towards laser fusion ignition. Nature **412**, 798 (2001)
7. Stacey, W.M., et al.: A TRU-Zr metal-fuel sodium-cooled fast subcritical advanced burner reactor. Nucl. Technol. **162**(1), 53 (2008)
8. Freidberg, J.P., Kadak, A.C.: Fusion-fission hybrids revisited. Nat. Phys. **5**, 370 (2009)
9. Singh, M.: Fail-Safe Source-Driven Fission and Fusion-Fission Hybrid Reactor Configurations. Master's thesis, University of Illinois, Urbana, Champaign (2013). https://www.ideals.illinois.edu/handle/2142/45583 and Ragheb, M., Singh, M.: Electrodynamic inertial fusion generators and fission satellites.

Thorium Energy Alliance 4th Annual Future of Energy Conference, Loyalla University, Water Tower Campus, Chicago, Illinois, USA (2012). Accessed 31 May–1 June 2012
10. Ragheb, M.M.H.: Fusio-fission thorium hybrid (2017). mragheb.com/…/Fission%20Fusion%20Thorium%20Hybrid.pdf
11. Ragheb, M.M.H., et al.: Nuclear performance of molten salt fusion-fission symbiotic systems for catalyzed deuterium-deuterium and deuterium-tritium reactors. Nucl. Tech. **48**, 217 (1980)
12. Rosenthal, M.W., et al.: The Development Status of Molten-Salt Breeder Reactors, ORNL-4812. Oak Ridge National Laboratory (1972). http://www.energyfromthorium.com/pdf/ORNL-4812.pdf
13. Briggs, R.B.: Molten-Salt Reactor Program Semi-Annual Progress Report For Period Ending 31 July, 1964. ORNL-3708, Oak Ridge National Laboratory (November 1964). http://www.energyfromthorium.com/pdf/ORNL-3708.pdf
14. Holcomb, D.E.: Molten Salt Reactors Today Status & Challenges, Workshop on MSR Technologies—Commemorating the 50th Anniversary of the Startup of the MSRE, ORNL, 15 Oct 2015
15. Shimazu, Y.: Current situation of MSR Development in Japan. Graduate School of Engineering, Hokkaido University, Japan (2007). http://www.uxc.com/smr/Library%5CDesign%20Specific/Fuji%20MSR/Presentations/2007%20-%20Current%20Situation%20of%20MSR%20Development%20in%20Japan.pdf
16. Garcia, H.E., et al.: Nuclear Hybrid Energy Systems—Regional Studies: West Texas & Northeastern Arizona: INUEXT-1 5-34503 (2015). http://www.inl.gov
17. Chen, J., et al.: Operations optimizations of nuclear hybrid energy systems. Nucl. Tech. **195**, 143 (2016)
18. Valenciaga, F., Puleston, P.F.: Supervisor control for a stand-alone hybrid generation system using wind and photovoltaic energy. IEEE Trans. on Energy Convers. **20**(2), 398 (2005)
19. Melicio, R., et al.: Wind Turbines with Permanent Magnet Synchronous Generator and Full-Power Converters: Modelling, Control and Simulation: Dr. Ibrahim Al-Bahadly (Ed.). ISBN: 978-953-307-221-0. InTech. http://www.intechopen.com/books/wind-turbines/wind-turbines-with-permanent-magnet-synchronousgenerator-and-full-power-converters-modelling-contro
20. Borowy, B.S., Salameh, Z.M.: Methodology for optimally sizing the combination of a battery bank and pv array in a wind/PV hybrid system: IEEE Trans. Energy Convers. **11**(2), 367–375 (1996)

3 Spallation Neutron Source, Multiplication and Possibility of Incineration

Spallation neutron source is new and has fastly attracted the attention of both nuclear and reactor scientists from the point of its application in incineration of the radioactive waste and nuclear energy production from the fertile fuels. This is a basic lesson of accelerator-driven subcritical system and future copious source of neutron for a variety of applications useful for the material science. Being higher energy than a fast reactor, secondary neutron built up in a reactor provides extra neutron flux required for double utilization in energy and nuclear transmutation. Simple kinetics of multiplication is discussed at large in the chapter. Following it CASCADE code ver. 04 calculation of neutron flux inside the IAEA benchmark design are performed for their application for the incineration of LLNW. The calculations have been compared with calculations of the MCNP code for the heat distribution.

3.1 Neutron Sources and Applications

Several kinds of neutron sources have been in existence, and they have a large number of applications including the sector of atomic energy. Production of a newer spallation neutron source is an application of accelerator technology. In case of nuclear energy, high-energy accelerator beam of charged particle with energy >30 MeV can produce a much harder neutron spectrum which can initiate fission in a fertile fuel which has hardly been burnt in a critical power reactor having neutron energy up to ~10.5 MeV. It may be mentioned that details of spallation neutron spectrum are the central theme of technology of energy amplifier for both energy and incineration of LLNW.

3.1.1 Neutron Sources

In 1930, Bothe and Becker [1] discovered a new radiation in bombardment of alpha radiation of polonium (Po) onto the beryllium target. They called new radiation to be γ radiation. In 1932, Chadwick identified it as neutron [2]. The following nuclear reaction (3.1) was found to be the cause of neutron production. Kinetic energy of alpha particle will contribute toward total kinetic energy of neutron which will escape out of the beryllium target.

$$ {}_2\mathrm{He}^4 + {}_4\mathrm{Be}^9 = {}_6\mathrm{C}^{12} + {}_0n^1 + Q(5.704\ \mathrm{MeV}) \tag{3.1}$$

Thus, it cannot be a unique value, and each neutron source will have its characteristic neutron energy spectrum. In Table 3.1, details of several neutron sources are given with data of neutron yield in each case. For example, if the strength of ^{241}Am/Be source is assigned to be 5Ci, this means, its rate of alpha emission is $5 \times 3.7 \times 10^{10}$/s and the corresponding neutron emission rate is $70 \times 5 \times 3.7 \times 10^{10}/$

V. K. Verma and K. Katovsky, *Spent Nuclear Fuel and Accelerator-Driven Subcritical Systems*, Green Energy and Technology, https://doi.org/10.1007/978-981-10-7503-2_3

Table 3.1 Details of the (α, n) sources. Alpha source, half-life, kinetic energy of alpha, neutron yield per 10^6 alpha particles, and percentage of neutrons with energy $E_n < 1.5$ MeV

Source	Half-life	E_α (MeV)	Neutron yield per 10^6 primary alpha particles		Percent yield with $E_n < 1.5$ MeV	
			Calculated	Experimental	Calculated	Experimental
^{239}Pu/Be	24,000 years	5.14	65	57	11	9–13
^{210}Po/Be	138 days	5.30	73	69	13	12
^{238}Pu/Be	87.4 years	5.48	79	–	–	–
^{241}Am/Be	433 years	5.48	82	70	14	–
^{244}Cm/Be	18 years	5.79	100	–	18	15–23
^{239}Cm/Be	162 days	6.10	118	106	22	29
^{239}Ra/Be	1,602 years	Multiple	502	–	26	26
^{227}Ac/Be	21.6 years	Multiple	502	–	28	33–38

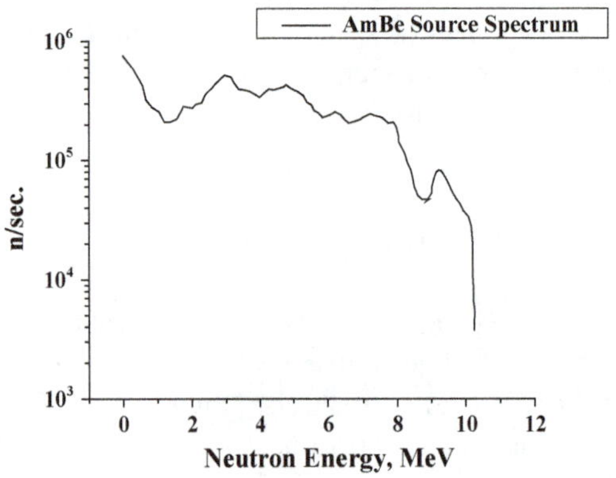

Fig. 3.1 Neutron energy spectrum of Am/Be source by Lorch [2]

$10^6 = 1.295 \times 10^7$ neutrons/s. In Fig. 3.1, neutron energy spectrum of Am/Be source by Lorch [2] is shown. This is one of the neutron source used for conduction of laboratory experiments. Also, the neutron sources given in Table 3.1 are frequently used in several deep underground geological searches of minerals, oil, and petroleum products as they are small in size and have long half-life in years. Similarly, ^{11}Be (α, n)^{14}N nuclear reaction occur with another isotope of Beryllium and in that case $Q = 0.158$ MeV.

Some of the nuclear reactions of neutron production useful in various research and developments can be briefly summarized as follows:

(d, n) reactions:

$$_1\text{H}^2 + {}_1\text{H}^3 = {}_2\text{He}^4 + n + 17.588\ \text{MeV} \quad (3.2)$$

$$_1\text{H}^2 + {}_3\text{Li}^7 = {}_4\text{Be}^8 + n + 15.028\ \text{MeV} \quad (3.3)$$

Again, these reactions can be performed in minimum laboratory conditions, and no acceleration of particles is required.

(p, n) reactions:

$$_1\text{H}^1 + {}_1\text{H}^3 = {}_2\text{He}^3 + n - 0.764\ \text{MeV} \quad (3.4)$$

$$_1\text{H}^1 + {}_3\text{Li}^7 = {}_4\text{Be}^7 + n - 1.646\ \text{MeV} \quad (3.5)$$

These reactions can be performed with the help of accelerated proton beams of kinetic energy greater than or equal to Q value.

(γ, n) reactions:

$$\gamma + {}_1\text{H}^2 = {}_1\text{H}^1 + n - 2.225\ \text{MeV} \quad (3.6)$$

$$\gamma + {}_4\text{Be}^9 = {}_4\text{Be}^8 + n - 1.666\ \text{MeV} \quad (3.7)$$

Reactions (3.6) and (3.7) are also called as 'nuclear photo-electric effect,' and they can be performed with the help of radioactive sources emitting gammas with energy, $E_\gamma > 1.666$ MeV. Also, neutrons can be produced in (γ, f) reactions with fissionable nuclei. Similarly, at high energies, (γ, $2n$) reactions are also possible.

In a fission reaction with any fuel element on an average, $\nu > 1$ neutrons are produced, and in a chain reaction in an infinitely large nuclear fuel, neutrons are multiplied. This results in large flux, $\sim 8.9 \times 10^{13}$ n/cm^2/s in a thermal reactor and $\sim 1.69 \times 10^{16}$ n/cm^2/s in a fast reactor.

3.1.2 Spallation Neutron Sources

Spallation as such is no word in English language, and it is derived from the verb spill of liquid from a container when some material is dropped in the liquid. This was the first time used in cosmic ray studies. After the early introduction of word spallation, nuclear reactions were reported at Berkeley in 1947 and the mechanism of spallation reactions was also discussed by Serber [3] in the same year. In collision of energetic particle like proton and deuteron having energy greater than several tens of MeV/n with a heavy nucleus, several neutrons are spilled out in early cascade along with pions and a part of other constituent nuclear matter of the nucleus. Subsequently, compound nucleus also decays emitting several nucleons and fragments. Fission process in case of actinides enhances the neutron production. The whole lot of neutrons produced in a high-energy collision are commonly identified as spallation neutrons, and they have high technological applications being energetic much beyond the energy of reactor neutrons other neutron sources as described in earlier Sect. 3.1.1. One of the applications of spallation neutrons has emerged in the form of 'energy amplifiers' [4, 5] for utilization of fertile uranium and thorium like fuels for nuclear energy and transmutation of the LLNW. The other application of spallation neutron source can be identified in the form of 'beam of cooled down neutrons' and separated in different energy ranges, hence different wavelengths as per their utilization in different applications. SINQ in Switzerland [6], ISIS in UK [7], SNS in USA [8], and JSNS in Japan [9] projects are already completed. The SNS project was one of the initiatives taken after the concept of EA [4] was proposed and its initial objective was to develop 1 GeV, 1 μs pulsed proton beam with 1 MW power. Its design details are discussed by Henderson et al. [10]. Similarly, JSNS with proton power of 1 MW, energy 3 GeV, current 333 mA, proton pulse time width 1 μs, frequency, 25 Hz is planned and their details will be discussed in Sect. 7.5. It is expected to produce $\sim 10^{17}$ n/s and a reactor of 15 MW power. For more details of the project, reader is advised to refer to [11].

Accelerated protons or heavier particles with energy >100 MeV/n on colliding with neutron-rich heavy targets like Hg, W, Pb, Bi, U, and Th produce large number of neutrons by different physical processes. Multiplicity of neutrons increases with kinetic energy of the projectile and length of the bulk of the target compared to the width of the target. An early review can be found in the article by Kumar et al. [12] wherein both experimental data of multiplicity has been compared with the model calculations. In Fig. 3.2(a), neutron multiplicity per incident proton (*n*/*p*) has been plotted to vary with length (cm) of the cylindrical lead target and compared with the calculated multiplicity from the CASCADE code version 2004 summarized in Ref. [12] itself. The two data show the following logarithmic growth of *n*/*p* with the length of the target.

Experimental data follows,

$$n/p = -8.76 + 8.44 \ln L \tag{3.8}$$

CASCADE code data follows,

$$n/p = -9.45 + 8.49 \ln L \tag{3.9}$$

In Fig. 3.2(b), the growth of neutron multiplicity with the diameter of the Pb target is plotted for proton projectile at 1.22 MeV where there are only two data points from the experiments [13]. Calculated data of the CASCADE code differs $\sim 9\%$ with the experiment data. From d = 8–30 cm, n/p growth is only 37% and it can be represented by the following exponential growth,

$$n/p = (25.64 \pm 3.43) - (13.53 \pm 1.85)\exp(-d/20.31) \tag{3.10}$$

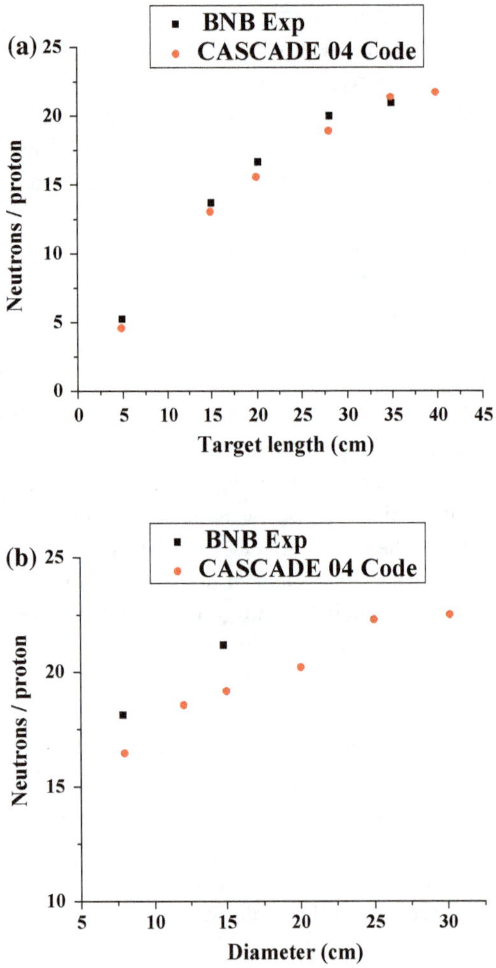

Fig. 3.2 Data of *n*/*p* from the BNB experiment [13] plotted as function of **(a)** length and **(b)** diameter of the lead (Pb) cylindrical target at 1.2 GeV energy of projectile proton and compared with results of calculations from the CASCADE 04 code [12]

Also, at BNL, neutron multiplicity is measured in reactions of proton colliding with W and Pb targets at energies ranging from 0.8 to 1.4 GeV [14], and the data is compared with calculated values obtained from the Monte Carlo CASCADE 04 code where better satisfactory results are found in case of Pb target [12].

In a Monte Carlo simulation study using the MCNPX code, Polański and Słowiński [15] have compared the neutron multiplicity and neutron spectra in proton and electron colliding with Pb target. In Fig. 3.3, neutron yield is plotted in case of electron beam energy colliding with the Pb target of size $d \times L = 3.4 \times 3.4$ cm^2. Neutron spectra of electron and proton beams colliding with Pb target of size $d \times L = 60 \times 60$ cm^2 are also plotted in Fig. 3.3. It may be pointed out that neutron spectra up to 1 MeV is similarly rising in both the cases but neutron multiplicity is about 50–100 times more in case of proton than the electron beam. At higher than 1 MeV neutron energy, it grows ~2 to ~4 orders of magnitude higher in case of protons than electron beam. On an average ratio, $n/e^- = 0.32$ and $n/p = 34$. On the contrary, activity induced by electron beam is found [15] to be 25 times higher than proton for the same target.

In another attempt, MC Simulation using MCNP-4C2 code shows that 5 MeV electron beam colliding with Be and BeD_2 targets [16] one can get moderated neutrons up to a flux of 1.23×10^8 n/cm^2/s/mA in the presence of graphite moderator. In case of electron, it is easier and economical to enhance beam current compared to proton acceleration. In Table 3.2, summary data of already existing spallation neutron sources is given.

Besides the aforesaid neutron sources [6–9], dedicated spallation neutron sources with several experimental facilities such as GAMMA [26, 27], energy and transmutation ($E + T$) [28–30], QUINTA [31], Subcritical Assembly at Dubna (SAD) [32], TARC [33], *n*-TOF [34] have used spallation neutron sources in basic experiments related to the ADS technology. Most of them are described in Chap. 6.

Using a Monte Carlo simulation code, such as CASCADE [35] and MCNP [36], neutron spectrum for a combination of the projectile, beam energy and target size can be calculated.

Spallation is the only process, known so far, of production of high flux of highly energetic neutrons from a small size target system. On the other hand, we know that from a thermal power reactor one can produce a neutron flux which is nearly two orders of magnitude smaller and energy is also limited to ~10.5 MeV compared to a spallation source that produces neutrons up to the projectile energy. The spallation targets are conveniently manageable for other applications; e.g., measurements of reaction cross sections

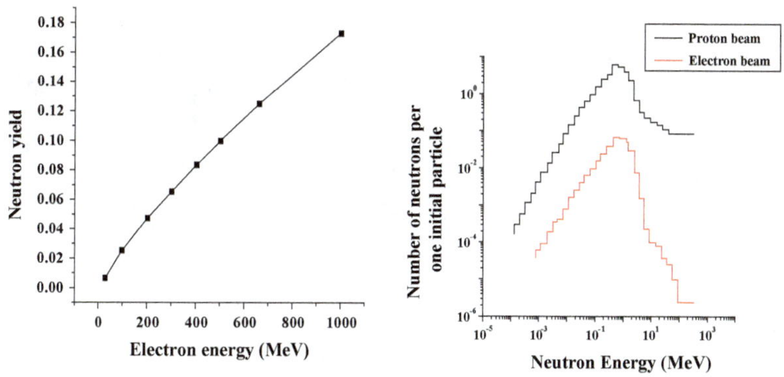

Fig. 3.3 (Left) Neutron yield plotted with electron energy from 30 to 1,000 MeV. (Right) neutron spectra in case of 1 GeV electron (lower histogram) and proton (upper histogram) beams colliding with Pb target calculated from the MCNP-4C2 code [15]

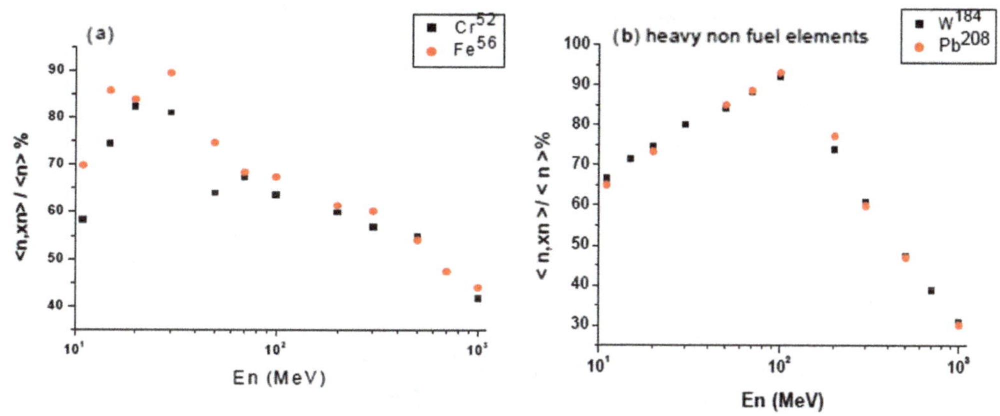

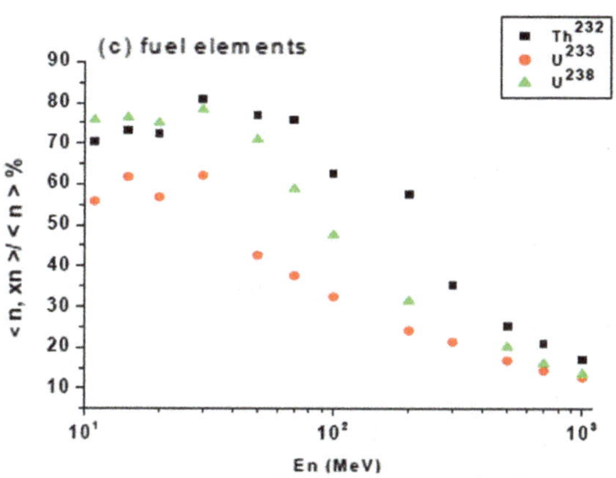

Data at 200 MeV for fuel

	< n >	< n_{xn}> / < n > %
Th^{232}	9.23	57.64
U^{238}	9.65	31.47
U^{233}	8.86	24.25

Fig. 3.4 Percentage contribution of (*n*, *xn*) nuclear process in neutron multiplication compared to total multiplicity plotted as a function of neutron energy E_n [50] for light mass materials, Cr and Fe, heavy elements, W and Pb, and fissionable fuel elements Th, U, and Pu. Data in the table corresponds to E_n = 200 MeV

[37–39] can be done at a fast rate using pulsed neutron source and production of beams of cold and ultra-cold neutrons [40] for a new area of neutron spectroscopy. Using high-energy neutrons for the measurement of cross section, *n*-ToF is the state-of-the-art technique because of

Table 3.2 Operation summary of high-energy spallation neutron sources

Name of the facility and location	Proton energy (MeV)/ Current (μA)	Beam power (kW)	Repetition rate (Hz)	Target material	Moderator	Year of facility operating or planned
LANSCE, Los Alamos National Laboratory (USA) [17]	800/70	56	20	Tungsten	L-H_2/ H_2O	1983
SNS, Oak Ridge National Laboratory (USA) [18, 19]	1,000/1,400	1,400	60	Mercury	L-H_2/ H_2O	2006
ISIS, Rutherford Appleton Laboratory (U.K.) [20, 21]	800/200	160	50	Tantalum	L-H_2/ H_2O	1985
IPNS, Argonne National Laboratory (USA) [22]	450/15	7	30	Depleted Uranium	S-CH_4/ L-CH_4	1981 (closed 2008)
SINQ, Paul Scherrer Institute, Switzerland [22, 23]	590/1,500	1,000	Continuous	Zircaloy	L-D_2/ D_2O	1996
JSNS, Japan Atomic Energy Agency, Japan [[24, 25]	3,000/333	1,000	25	Mercury	L-H_2	2008

its high resolution. On a big scale, the *n*-ToF facilities are working at CERN [34, 37–39], GEANIE facility and LANCE [41]. Other facilities using neutrons with nearly reactor energy for cross-section measurements are cyclotron CYCLONE at Louvain-la-Neuve [42, 43], Pohang facility using (γ, n) source [44], and GELINA at Geel [45]. Intense resonance neutron source (IREN) [46] has also started working at JINR, Dubna.

For designing of an ADSS as energy and transmutation system, basic data of neutron microscopic cross sections at energies higher than 20 MeV is rarely available and this will take several years to get such data for a large number of elements and several reactions. Using the low-intensity spallation sources [26–31] at JINR Dubna, several attempts have been made to measure spectrum average cross sections (sp. av. cs.) of both fissionable materials and other structural materials of ADS [27, 28, 30, 31, 47, 48]. The sp. av. cs. are derived from the measurements of reaction rates measured in an experiment.

3.2 Neutron Multiplicative Processes

In case of high-energy neutron interaction with a nucleus, there can be scattering, catastrophic inelastic reaction or a complete absorption and formation of a compound nucleus. Different to complete absorption, a slow neutron on absorption by a fissionable nucleus, there is good chance that net neutron balance is positive. In case of an ADSS, there is copious supply of high-energy spallation neutrons, and multiplication of neutrons will not only be from the fuel elements but also from other reactor materials. This makes ADSS a quite different reactor. Also, in presence of enormous yield of neutrons, there will be much higher radiation damage of the ADSS reactor material compared to normal critical power reactor. Whether it is a fissionable material or other structure material, the following neutron multiplicative processes are identified for considerations:

(i) Single-neutron type—(*n*, *n*′), (*n*, *np*), (*n*, *nd*), (*n*, *nt*), (*n*, *nα*), (*n*, nHe^3) and (*n*, *n2p*), etc.
(ii) Multiple-neutron type—(*n*, *xn*) where $x = 2, 3, \ldots,$ (*n*, *f*), (*n*, $2n$He^3), (*n*, *2nd*), (*n*, *2nt*), (*n*, *2npd*), (*n*, *2n2p*), (*n*, *3np*), (*n*, *3nd*), (*n*, $3n$He^3), (*n*, *3nt*) and (*n*, *3nα*), etc.
(iii) Neutron removal type—(*n*, γ), (*n*, *f*), (*n*, *p*), (*n*, *d*), (*n*, *t*) and (*n*, α), etc.

As mentioned earlier, neutron multiplication by (a) structure material used in the reactor, i.e., spallation target, moderator, fuel carrying tubes or fuel cartridges and the shielding material and (b) fuel elements are different processes and they need separate analysis. In fact, it adds a dimension to the neutronics of a reactor.

3.2.1 Role of (*n*, *xn*) and Similar Reactions

In case of any of the Pb, Bi, Pb–Bi eutectic, or any other material of high atomic number, the (*n*, *xn*) and (*n*, *xnyp*)-type reactions are important due to multiplication of neutrons by the produced high-energy neutrons. Secondly, even in case of fuel elements like U, Th, or Pu, neutrons are added by way of (*n*, *xn*), higher than $x = 2$ order and (*n*, *xnyp*)-type reactions to the fission neutrons. This makes ADSS different than critical reactors [12] because its fuel cycles are extended enormously. Thirdly, due to higher energy radiation, heavy radiation damage takes place, and this stresses upon need of development of radiation resistant materials and to extract highly radiotoxic materials frequently during the operation. In this way, having availability of additional neutrons beyond the energy requirement, incineration of long-lived isotopes like iodine (I), technetium (Tc), and plutonium (Pu) can be possible. Lastly, contribution of (*n*, γ) reactions in the process of transmutation will rather be reduced because of smaller cross section of the reaction compared to a critical reactor where the neutron energy spectrum is relatively smaller. For details, reader may refer [49–51].

As the list of materials used in construction of a hybrid ADSS reactor is expected to be large and different to the existing critical reactors, therefore, there is need of additional new experimental data with higher accuracy. Presently, for the development of a design of a prototype after developing understanding of (*n*, *xn*)-type reactions, new data libraries are being prepared using the simulated data. Using codes like TALYS-1.0 to calculate cross sections up to 250 MeV energy [52] and CASCADE.04 code in the range from 11 to 1,000 MeV energy [53], for a large number of materials data tables are prepared and presented in reference [49]. In case of several reactions, experimental data [54–56] have been compared with the calculated data from the codes. In case of some of the ADSS materials, percentage contributions of (*n*, *xn*) and (*n*, *f*) reactions are assessed based on the calculated data and presented in Table 3.3.

The detailed calculated data of neutron multiplicity ratio, $\langle n\ (n,\ xn)\rangle / \langle n\rangle$, is plotted in Fig. 3.4 as a function of spallation neutron energy, E_n (MeV), [49] up to 1,000 MeV. The following inferences can be drawn from the data,

(i) Light structure materials of different masses, e.g., Cr^{52} and Fe^{56}, follow a trend different to heavy structure materials like ^{184}W and ^{208}Pb. Heavy materials show a unique rise of the ratio, $\langle n(n,\ xn)\rangle / \langle n\rangle$ up to 100 MeV and at higher energies it declines.
(ii) The ratio, $\langle n(n,\ xn)\rangle / \langle n\rangle$, when plotted in case of ^{232}Th, ^{238}U, and ^{233}U shows no unique feature like heavy structure materials discussed above. Rather the ratio follows, $^{232}\text{Th} > {}^{238}\text{U} > {}^{233}\text{U}$. It can be inferred that the contribution of (*n*, *xn*) process in neutron multiplication is more in case of more fertile nuclei and least in case of fissile nucleus.

Table 3.3 Total sum of cross sections of (n, xn) reactions, $\Sigma\sigma(n, xn)$ for all x values at the 200 MeV neutron energy for several ADSS materials, fission cross section $\Sigma\sigma(n, f)$ for ^{238}U, ^{235}U, ^{233}U, and ^{232}Th in barns (b) and their percentage contribution w.r.t. total non-elastic cross sections calculated using TALYS-1.0

Element	$\Sigma\sigma(n, xn)$ (mb)	$\Sigma\sigma(n, f)$ (mb)	$\Sigma\sigma_{\text{non-elastic}}$ (b)	$\frac{\sum\sigma(n,xn)}{\sum\sigma_{\text{non-elastic}}}\%$	$\frac{\sum\sigma(n,f)}{\sum\sigma_{\text{non-elastic}}}\%$	$\frac{\sum\sigma(n,xn)}{\sum\sigma(n,f)}$
^{238}U	273.32 *1,846.54	811.02 *262.46	2.11 *2.98	12.96 *62.04	38.46 *8.82	0.34 *7.03
^{235}U	249.28 *1,886.83	587.57 *677.97	1.93 *3.20	12.90 *59.01	30.42 *21.20	0.42 *2.78
^{233}U	272.85 *581.20	538.80 *1,989.7	1.92 *3.21	14.22 *18.12	28.07 *62.02	0.51 *0.29
^{232}Th	357.53 *1,880.13	320.55 *86.71	2.08 *2.93	17.17 *64.12	15.38 *2.96	1.12 *21.66
^{209}Bi	535.81	–	1.78	30.18	–	–
^{208}Pb	495.71	–	1.77	28.04	–	–
^{184}W	406.72	–	1.64	24.88	–	–
^{181}Ta	410.08	–	1.60	25.65	–	–
^{96}Mo	90.06	–	0.97	9.25	–	–
^{93}Nb	71.77	–	0.94	7.66	–	–
^{90}Zr	65.53	–	0.92	7.10	–	–
^{65}Cu	93.80	–	0.72	13.08	–	–
^{63}Cu	37.80	–	0.70	5.37	–	–
^{59}Co	23.77	–	0.62	3.82	–	–
^{58}Ni	4.66	–	0.62	0.75	–	–
^{56}Fe	17.83	–	0.59	3.04	–	–
^{55}Mn	27.29	–	0.58	4.68	–	–
^{52}Cr	22.95	–	0.57	4.06	–	–
^{51}V	28.77	–	0.54	5.37	–	–
^{48}Ti	22.42	–	0.53	4.23	–	–
^{31}P	1.46	–	0.38	0.38	–	–
^{28}Si	1.44	–	0.36	0.40	–	–
^{27}Al	3.27	–	0.33	0.99	–	–

In the last column ratio, $\Sigma\sigma(n, xn)/\Sigma\sigma(n, f)$ is given. All the data corresponds to $E_n = 200$ MeV. Data with (*) corresponds to 10 MeV energy for comparing with data of 200 MeV

Among the considered elements and energy, $E_n = 200$ MeV, the ratio $\frac{\Sigma\sigma(n,xn)}{\Sigma\sigma_{\text{nonelas}}}\%$ is highest 30.18% for ^{209}Bi nuclide followed by ^{208}Pb, ^{181}Ta, and ^{184}W, where the ratio is 28.04, 25.55, and 24.84, respectively. Likewise, the ratio for fuel elements ^{238}U, ^{235}U, ^{233}U, and ^{232}Th is 12.96, 12.90, 14.21, and 17.17, respectively. From the neutron multiplication point of view, ^{209}Bi is very good. The ratio $\frac{\Sigma\sigma(n,xn)}{\Sigma\sigma(n,f)}\%$ is an indication of a competition between (n, xn) and (n, f) reactions, and its calculated values for ^{238}U, ^{235}U, ^{233}U, and ^{232}Th at 200 MeV are 0.337, 0.424, 0.506, and 1.12, respectively. This clearly indicates that although the (n, xn) reactions are not so important compared to (n, f) reactions in case of uranium, they play a vital role in the case of ^{232}Th fuel. For lower values of energy, e.g., at 10 MeV, the values of ratio $\frac{\Sigma\sigma(n,xn)}{\Sigma\sigma(n,f)}\%$ for ^{238}U, ^{235}U, ^{233}U, and ^{232}Th are 7.03, 2.78, 0.29, and

21.66%, respectively. Thus, at 10 MeV energy, the $(n, 2n)$ reactions are dominant in case of ^{232}Th compared to ^{238}U and ^{235}U also. The ratio can be implemented to define fissional character of a fuel. Thus, the fissional order of four elements can be written as $^{233}\text{U} > {}^{235}\text{U} > {}^{238}\text{U} > {}^{232}\text{Th}$.

3.2.2 Multiplication Coefficient and Source Importance

Let us assume that a cylindrical spallation neutron source of size $d \times L = 20 \times 50$ cm^2 is enclosed inside another cylindrical of x cm thickness and filled with a fuel. In Fig. 3.5, two analogous cylindrical structures (a) and (b) are shown. In the inner cylinder of Fig. 3.5(a), a static neutron source is assumed, and in case of Fig. 3.5(b), a spallation source irradiated by 1 GeV proton beam of 1 A current is shown.

Spallation neutron energy spectrum produced in 1,000 MeV proton collision with massive lead (Pb) target of size $d \times L = 20 \times 50$ cm^2 is calculated from the CASCADE 04 code [51], and it has been plotted in Fig. 3.6. The spectrum lies between 20 eV to ~940 MeV compared to a thermal and a fast reactor spectrum which shows maximum neutron energy to be ~10.5 MeV [57]. Spectrum average cross section (sp. av. cs.) defined [58] over the range of a neutron spectrum can be calculated from the flat or group cross sections from the raw cross section data at individual energy. In Table 3.4, calculated values of sp. av. cs. of various reactions occurring in ^{232}Th, ^{233}U, ^{235}U, and ^{238}U fuel elements by the three neutron spectra of thermal, fast reactors and the spallation spectrum are given.

According to a simple derivation for the neutron multiplication by Cullen [59], when I_0 intensity of incident neutrons is allowed to fall on a material in a given direction, then the intensity of surviving neutrons after passing through distance 'x' may be given by,

$$I_x = I_0 \exp(-x\Sigma_t) \qquad (3.11)$$

where Σ_t is the total macroscopic cross section of a neutron in the given material. The formulation is applicable as the neutron is non-ionizing, and it changes direction on interaction only. Intensity of the neutron that interacts in a traversal of distance x may be written as follows:

$$(I_0 - I_x) = I_{\text{int}} = I_0(1 - \exp(-x\Sigma_t)) \qquad (3.12)$$

The probability of interaction, $P_1 = I_{\text{int}}/I_0$ and the interaction can be of any of the elastic, capture, or fission, etc. Thus, probability of a type of

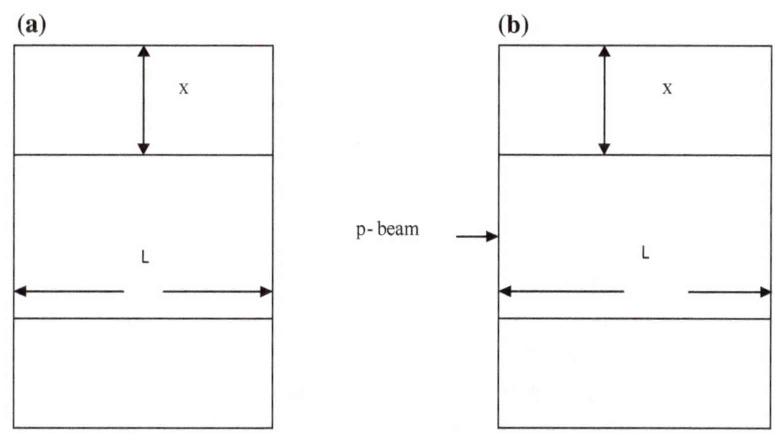

Fig. 3.5 A double cylindrical structure is used in calculations: (a) Inner cylinder is assumed as the source of neutrons, and the outer cylinder is filling with fuel element, (b) inner cylinder of lead is a spallation target irradiated with 1 GeV proton. Thickness of outer cylinder is assumed to vary as per requirement of neutron multiplication

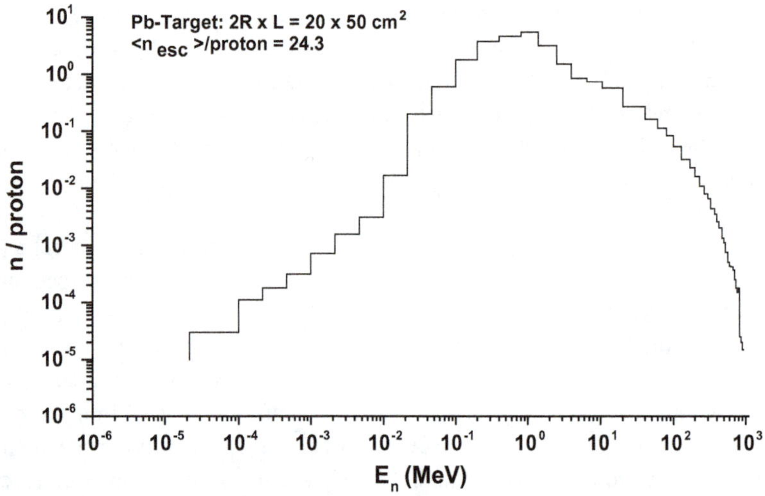

Fig. 3.6 Spallation neutron spectrum of 1 GeV proton colliding with massive Pb target calculated from the CASCADE 04 code [51]. There are 0.3% neutrons with E_n > 250 MeV

Table 3.4 Spectrum average cross sections, σ (mb) for different reaction channels of the fertile ^{232}Th and ^{238}U and fissile ^{233}U and ^{235}U fuel elements for the spallation, fast, and thermal neutron fluxes. Here, symbols correspond to standard reaction channels of TALYS-1.0 code [52]

$\langle\sigma$ (reactions)$\rangle$	^{232}Th	^{233}U	^{235}U	^{238}U
	Spallation Fast Thermal	**Spallation** Fast Thermal	**Spallation** Fast Thermal	**Spallation** Fast Thermal
$\langle\sigma$ $(n, \gamma)\rangle$	**146.1** 391 5,350	**41.8** 74.2 431	**106.4** 270 1,730	**116.8** 317 5,490
$\langle\sigma$ $(n, n')\rangle$	**2,022.7** 856 707	**1,102.4** 470 405	**1,684.7** 835 642	**2,632.2** 1,510 995
$\langle\sigma$ $(n, 2n)\rangle$	**74.8** 7.85 8.97	**27.7** 3.17 3.97	**89.2** 15.3 21.6	**70.8** 13.7 17.4
$\langle\sigma$ $(n, f)\rangle$	**35.1** 7.29 8.46	**2,209.7** 2,250 12,600	**1,184.2** 2,080 12,900	**72.4** 11.4 14.6
$\langle\sigma$ $(n, 3n)\rangle$	**41.38**	**12.97**	**32.19**	**41.20**
$\langle\sigma$ $(n, 4n)\rangle$	**8.98**	**6.58**	**3.53**	**6.86**
$\langle\sigma$ $(n, 5n)\rangle$	**12.11**	**11.51**	**6.89**	**7.74**
$\langle\sigma$ $(n, 6n)\rangle$	**6.98**	**3.42**	**4.35**	**8.13**
$\langle\sigma$ $(n, 7n)\rangle$	**6.97**	**1.43**	**1.35**	**1.71**
$\langle\sigma$ $(n, 8n)\rangle$	**0.86**	**0.83**	**1.46**	**0.81**
$\langle\sigma$ $(n, 9n)\rangle$	**0.56**	**0.67**	**0.89**	**0.82**
$\langle\sigma$ $(n, 0n)\rangle$	**0.54**	**0.47**	**0.62**	**0.81**
$\langle\sigma$ $(n, 11n)\rangle$	**0.63**	**0.20**	**0.09**	**0.79**
$\langle\sigma$ $(n, 12n)\rangle$	**0.06**	**0.03**	**0.20**	**0.07**
$\langle\sigma$ $(n, 13n)\rangle$	**0.11**	**0.12**	**0.13**	**0.07**
$\langle\sigma$ $(n, 14n)\rangle$	**0.06**	**0.0523**	**0.0667**	**0.0209**
$\langle\sigma$ $(n, 15n)\rangle$	**0.034**	**0.026**	**0.0332**	**0.0107**

(continued)

Table 3.4 (continued)

⟨σ (reactions)⟩	^{232}Th	^{233}U	^{235}U	^{238}U
	Spallation Fast Thermal	**Spallation** Fast Thermal	**Spallation** Fast Thermal	**Spallation** Fast Thermal
⟨σ (*n*, *p*)⟩	**0.371**	**0.51**	**0.42**	**0.35**
⟨σ (*n*, *d*)⟩	**0.19**	**0.27**	**0.23**	**0.18**
⟨σ (*n*, *t*)⟩	**0.05**	**0.11**	**0.09**	**0.06**
⟨σ (*n*, He^3)⟩	**1.17 × 10^{-4}**	**1.25 × 10^{-4}**	**9.93 × 10^{-5}**	**8.30 × 10^{-5}**
⟨σ (*n*, α)⟩	**0.05**	**0.24**	**0.17**	**0.05**
⟨σ (*n*, *np*)⟩	**1.15**	**1.51**	**1.33**	**1.07**
⟨σ (*n*, *nd*)⟩	**0.17**	**0.18**	**0.16**	**0.16**
⟨σ (*n*, *nt*)⟩	**0.06**	**0.09**	**0.08**	**0.05**
⟨σ (*n*, *n*α)⟩	**0.04**	**0.14**	**0.03**	**0.04**
⟨σ (*n*, *n*He^3)⟩	**2.88 × 10^{-3}**	**5.64 × 10^{-3}**	**0.00487**	**3.35 × 10^{-3}**
⟨σ (*n*, *pd*)⟩	**1.26 × 10^{-4}**	**2.54 × 10^{-4}**	**1.99 × 10^{-4}**	**1.84 × 10^{-4}**
⟨σ (*n*, *npd*)⟩	**1.68 × 10^{-4}**	**7.23 × 10^{-4}**	**6.68 × 10^{-4}**	**2.74 × 10^{-4}**
⟨σ (*n*, 2*p*)⟩	**4.93 × 10^{-4}**	**3.09 × 10^{-4}**	**3.33 × 10^{-4}**	**4.73 × 10^{-5}**
⟨σ (*n*, 2*n*He^3)⟩	**4.98 × 10^{-3}**	**8.49 × 10^{-3}**	**5.25 × 10^{-3}**	**3.92 × 10^{-3}**
⟨σ (*n*, 2*nd*)⟩	**0.18**	**0.25**	**0.23**	**0.14**
⟨σ (*n*, 2*nt*)⟩	**0.11**	**0.14**	**0.12**	**0.10**
⟨σ (*n*, 2*n*α)⟩	**1.46 × 10^{-3}**	**3.86 × 10^{-3}**	**1.72 × 10^{-3}**	**6.91 × 10^{-4}**
⟨σ (*n*, 2*npd*)⟩	**8.40 × 10^{-4}**	**1.57 × 10^{-3}**	**2.56 × 10^{-3}**	**7.64 × 10^{-4}**
⟨σ (*n*, *n*2*p*)⟩	**4.31 × 10^{-3}**	**5.62.10^{-3}**	**3.48 × 10^{-3}**	**3.31 × 10^{-3}**
⟨σ (*n*, 2*np*)⟩	**0.44**	**0.39**	**0.39**	**0.40**
⟨σ (*n*, 2*n*2*p*)⟩	**4.15 × 10^{-3}**	**0.0115**	**9.27 × 10^{-3}**	**4.86 × 10^{-3}**
⟨σ (*n*, 3*np*)⟩	**1.39**	**1.56**	**1.56**	**0.99**
⟨σ (*n*, 3*nd*)⟩	**0.34**	**0.40**	**0.34**	**0.25**
⟨σ (*n*, 3*n*He^3)⟩	**3.11 × 10^{-3}**	**6.60 × 10^{-3}**	**4.85 × 10^{-3}**	**2.14 × 10^{-3}**
⟨σ (*n*, 3*nt*)⟩	**0.35**	**0.38**	**0.16**	**0.18**
⟨σ (*n*, 3*n*α)⟩	**0.15**	**0.43**	**0.30**	**0.11**

interaction can be written as $P_2 = \Sigma/\Sigma_t$ where Σ is the macroscopic cross section of a type of interaction. Thus, for $I_0 = 1$, neutron intensity, fraction probability of a particular type of interaction may be written as:

$$P_1 \cdot P_2 = (1 - \exp(-x\Sigma_t))(\Sigma/\Sigma_t) \quad (3.13)$$

Thus, the production of neutrons in the material filled in x thickness of the outer cylinder of Fig. 3.5 can be written as:

$$\begin{aligned} P = {} & (1 - \exp(-x\Sigma_t)) \cdot [2\Sigma(n, 2n) + 3\Sigma(n, 3n) \\ & + 4\Sigma(n, 4n) \cdots + 9\Sigma(n, 9n) \cdots \\ & + \langle v \rangle \Sigma(n, f) + 2\Sigma(n, 2np) \cdots \\ & + 3\Sigma(n, 3n\alpha) + \cdots]/\Sigma_t \end{aligned} \quad (3.14)$$

Here, $\langle v \rangle$ is the average number of fission neutrons, and this depends on the fuel element.

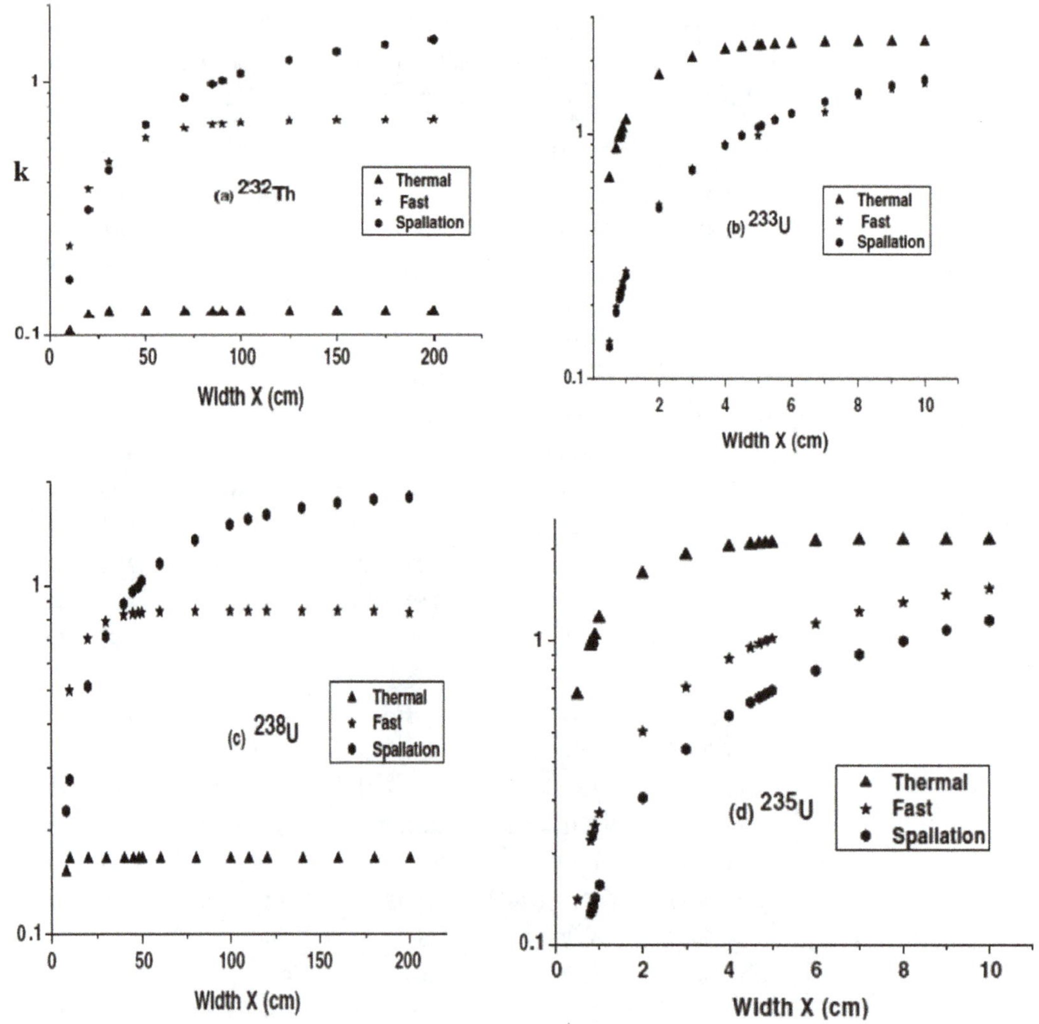

Fig. 3.7 Variation of neutron multiplication coefficient k with the thickness X (cm) for the thermal, fast, and the spallation neutron spectra for **(a)** ^{232}Th, **(b)** ^{233}U, **(c)** ^{238}U, and **(d)** ^{235}U fuel elements when irradiated by the thermal, fast, or spallation neutron spectra [60]

Similarly, removal term, R, for the process of removal, absorption, or utilization of neutrons in the reaction taking place in x thickness can be written as,

$$\begin{aligned} R = {} & (1-\exp(-x\Sigma_t))\cdot[(\Sigma(n,\gamma)+\Sigma(n,p) \\ & +\Sigma(n,d)+\cdots+\Sigma(n,np) \\ & +\Sigma(n,2n)\cdots+\Sigma(n,9n)+\cdots+\Sigma(n,f) \\ & +\Sigma(n,2np)\cdots+\Sigma(n,3n\alpha)\ldots]/\Sigma_t \end{aligned} \tag{3.15}$$

In this analysis, only those reactions are accounted that are initiated by neutrons. Other reactions that can contribute substantially are (γ, n) and (γ, f)-type reactions are avoided. Thus, the neutron multiplication coefficient k can be written as:

$$k = P/(R+L) \tag{3.16}$$

Here, L is the neutron loss in leakage. For incident neutron intensity to be unity, i.e., $R + L = 1$, the multiplication coefficient k can be

Table 3.5 Spectrum average cross sections for the neutron poisons in the given thermal, fast, and spallation neutron fluxes

Neutron poison	Spectrum average cross section		
	Thermal spectrum (b)	Fast spectrum (mb)	Spallation spectrum (mb)
^{83}Kr (n, γ)	48.46	205	29.29
^{95}Mo (n, γ)	5.80	247	47.27
^{99}Tc (n, γ)	15.25	514	78.29
^{101}Ru (n, γ)	3.12	567	68.26
^{103}Rh (n, γ)	51.35	479	72.68
^{105}Pd (n, γ)	6.33	719	104.62
^{107}Pd (n, γ)	2.93	762	119.11
^{133}Cs (n, γ)	19.15	381	56.34
^{135}Xe (n, γ)	8.19×10^5	52.5	10.09
^{143}Nd (n, γ)	76.32	214	45.55
^{147}Pm (n, γ)	103.12	936	153.69
^{149}Sm (n, γ)	25,070.44	1,600	199.86

calculated [60] from the value of P. For the three isotopes of uranium and ^{232}Th on filling in the outer cylinder of design of Fig. 3.5(b) one by one in different thicknesses, x results are presented in Fig. 3.7(a)–(d) for ^{232}Th, ^{233}U, ^{238}U, and ^{235}U, respectively. From this, the following inferences can be drawn,

(i) Both the fertile fuels ^{232}Th and ^{238}U reach a critical limit, $k = 1$, in case of spallation neutrons, and the required fuel thickness comes out to be ~100 cm. In a fast reactor, ^{232}Th fuel attains a maximum value, $k = 0.5$, while ^{238}U attains $k = 0.8$.
(ii) In case of fissile fuels ^{233}U and ^{235}U, both fast and spallation neutron spectra work similarly for attaining a value $k = 1$ at ~6 cm. Thermal flux is better for early attainment of $k > 1$ value than fast and spallation spectra.
(iii) In case of ^{235}U, multiplication coefficient k attains a value differently to ^{233}U in the three neutron spectra. At a given thickness, X value of k follows: $k_{\text{Therm}} > k_{\text{Fast}} > k_{\text{Spall.}}$

In critical reactors, nuclides ^{133}Cs, ^{101}Ru, ^{103}Rh, ^{99}Tc, ^{105}Pd, ^{107}Pd, and ^{149}Sm are treated as the neutron poisons in the lead-bismuth cooled fast breeder reactor [61] and ^{83}Kr, ^{95}Mo, ^{143}Nd, ^{147}Pm, ^{135}Xe, and ^{149}Sm in the thermal flux environment [62]. For the spectrum of spallation neutrons given in Fig. 3.6, the sp. av. cs. are calculated [60] neutron poisons and presented in the last column of Table 3.5 along with the thermal and fast spectra. From the data, it may be inferred that the given neutron poisons of thermal and fast reactors are least effective in case of spallation neutrons and their neutron absorption cross sections can be put in the following order:

$$\sigma^{\text{sp.av.cs.}}(\text{thermal}) > \sigma^{\text{sp.av.cs.}}(\text{fast}) > \sigma^{\text{sp.av.cs.}}(\text{spallation})$$

3.3 Utilization of Spallation Neutrons by a Fuel

Utilization of spallation neutrons is better visualized [60] by imparting neutron on a big block of ^{232}Th, say, with $X = 81.7$ cm and similarly on a block of fissile ^{233}U, say, with $X = 4.5$ cm. Calculated results of contributions of various reactions to the P and R terms using Eqs. (3.14) and (3.15) are presented in Table 3.6.

Neutron multiplication by way of (n, f) reaction in ^{232}Th corresponding to $X = 81.7$ cm thickness of the outer fuel cylinder of Fig. 3.6 is 0.150/0.982 = 15.27% per incident spallation neutron. This also means that spallation neutrons

Table 3.6 Details of removal (R) and production (P) per incident neutron for ^{232}Th and ^{233}U fuel elements filled in the two different thicknesses 'X' of cylinder independent to each other

Channel	Removal, R	Production, P	Removal, R	Production, P
	^{232}Th (X = 81.7 cm)		^{233}U (X = 4.5 cm)	
(n, γ)	0.251	–	7.14×10^{-3}	–
(n, 2n)	0.128	0.257	4.73×10^{-3}	9.46×10^{-3}
(n, 3n)	7.10×10^{-2}	0.2129	2.21×10^{-3}	6.64×10^{-3}
(n, 4n)	1.54×10^{-2}	6.16×10^{-2}	1.12×10^{-3}	4.49×10^{-3}
(n, 5n)	2.08×10^{-2}	0.104	1.96×10^{-3}	9.82×10^{-3}
(n, 6n)	1.20×10^{-2}	7.18×10^{-2}	5.84×10^{-4}	3.50×10^{-3}
(n, 7n)	1.20×10^{-2}	8.37×10^{-2}	2.44×10^{-4}	1.71×10^{-3}
(n, 8n)	1.47×10^{-3}	1.17×10^{-2}	1.42×10^{-4}	1.13×10^{-3}
(n, 9n)	9.50×10^{-4}	8.60×10^{-3}	1.14×10^{-4}	1.02×10^{-3}
(n, 10n)	9.02×10^{-4}	9.02×10^{-3}	7.99×10^{-5}	7.99×10^{-4}
(n, 11n)	1.05×10^{-3}	1.15×10^{-2}	3.43×10^{-5}	3.77×10^{-4}
(n, 12n)	1.07×10^{-4}	1.28×10^{-3}	4.56×10^{-6}	5.47×10^{-5}
(n, 13n)	1.84×10^{-4}	2.38×10^{-3}	1.92×10^{-5}	2.49×10^{-4}
(n, 14n)	9.85×10^{-5}	1.38×10^{-3}	8.93×10^{-6}	1.25×10^{-4}
(n, 15n)	5.74×10^{-5}	8.60×10^{-4}	4.35×10^{-6}	6.53×10^{-5}
(n, p)	6.40×10^{-4}	–	8.70×10^{-5}	–
(n, d)	3.26×10^{-4}	–	4.58×10^{-5}	–
(n, t)	8.47×10^{-5}	–	1.72×10^{-5}	–
(n, He^3)	1.95×10^{-7}	–	2.13×10^{-8}	–
(n, α)	8.14×10^{-5}	–	4.01×10^{-5}	–
(n, f)	6.01×10^{-2}	0.150	0.377	0.940
(n, np)	2.59×10^{-3}	2.60×10^{-3}	2.58×10^{-4}	$2.60.10^{-4}$
(n, nd)	2.95×10^{-4}	2.95×10^{-4}	3.01×10^{-5}	3.01×10^{-5}
(n, nt)	1.01×10^{-4}	1.01×10^{-4}	1.58×10^{-5}	1.58×10^{-5}
(n, nHe^3)	4.95×10^{-6}	4.94×10^{-6}	9.63×10^{-7}	9.63×10^{-7}
(n, na)	7.60×10^{-5}	7.60×10^{-5}	2.34×10^{-5}	2.33×10^{-5}
(n, 2p)	8.23×10^{-7}	–	5.27×10^{-8}	–
(n, pd)	2.10×10^{-7}	–	4.33×10^{-8}	–
(n, $n2p$)	7.39×10^{-6}	7.39×10^{-6}	9.59×10^{-7}	9.59×10^{-7}
(n, 2np)	2.48×10^{-3}	4.97×10^{-3}	2.27×10^{-4}	4.53×10^{-4}
(n, 2na)	8.08×10^{-5}	1.62×10^{-4}	2.08×10^{-5}	4.16×10^{-5}
(n, 2nd)	3.09×10^{-4}	6.18×10^{-4}	4.26×10^{-5}	8.53×10^{-5}
(n, 2nt)	1.78×10^{-4}	3.55×10^{-4}	2.41×10^{-5}	4.81×10^{-5}
(n, 2$n\alpha$)	8.55×10^{-6}	1.71×10^{-5}	1.45×10^{-6}	2.90×10^{-6}
(n, 2npd)	1.44×10^{-6}	2.88×10^{-6}	2.69×10^{-7}	5.37×10^{-7}
(n, 2n2p)	7.12×10^{-6}	1.42×10^{-5}	1.97×10^{-6}	3.94×10^{-6}
(n, 3np)	2.38×10^{-3}	7.13×10^{-3}	2.66×10^{-4}	7.98×10^{-4}
(n, 3nd)	5.81×10^{-4}	1.74×10^{-3}	6.85×10^{-5}	2.06×10^{-4}
(n, 3nt)	6.05×10^{-4}	1.82×10^{-3}	6.39×10^{-5}	1.92×10^{-4}

(continued)

Table 3.6 (continued)

Channel	Removal, R	Production, P	Removal, R	Production, P
	^{232}Th (X = 81.7 cm)		^{233}U (X = 4.5 cm)	
(n, 3nHe^3)	5.33×10^{-6}	1.60×10^{-5}	1.13×10^{-6}	3.38×10^{-6}
(n, 3na)	2.57×10^{-4}	7.71×10^{-4}	7.32×10^{-5}	2.19×10^{-4}
Sum	0.684	0.981	0.397	0.981
k_{eff}	0.982		0.981	
No. of atoms (mass)	4.26×10^{28} (16.39 ton)		8.33×10^{26} (0.32 ton)	

generate nearly 84.73% more neutrons by the reactions other than fission processes, and these processes are not so important in case of fissile materials. In case of ^{233}U for thickness, $X = 4.5$ cm, contribution of (n, f) process is 95.82% per incident spallation neutron and other nuclear processes contribute only 4.18% toward neutron multiplication. In the following, a simple way of calculation of fission energy by the spallation neutrons is presented.

Energy Production in 81.7 cm Thick ^{232}Th
According to CASCADE code ver. 2004, in case of 1,000 MeV proton colliding with Pb target of size $d \times L = 20 \times 50$ cm^2, on an average 24.3 spallation neutrons/p are expected to be produced. Thus, in passing through 81.7 cm thickness of ^{232}Th, there are:

(i) Applied neutrons/p = $0.982 \times 24.3 = 23.86$ n/p
(ii) Neutrons used for fission = $23.86 \times 0.15 = 3.58$ n/p
(iii) Neutrons, not involved in fission = 23.86–3.58 = 20.28/p
(iv) Fission energy production, $H_{0.232} = 3.58 \times 186.7 = 669.3$ MeV/p

Energy Production in 4.5 cm thick ^{233}U

(i) Neutrons used for fission = $23.86 \times 0.94 = 22.43$ n/p
(ii) Neutrons, not involved in fission = 1.43/p
(iii) Fission energy produced, $H_{0.233} = 22.43 \times 197.9 = 4{,}438.6$ MeV/p

This is ~4.4 times higher than input energy 1,000 MeV/p incident energy.

Let us assume that all of the P-neutrons involved in fission is multiplied in each subsequent cascade process over an infinite volume of the subsequent fuel as does in a reactor; then, energy production may be expressed as,

$$H = H_0\left(\frac{k_{\text{eff}}}{1 - k_{\text{eff}}}\right) \tag{3.17}$$

Thus, the ^{232}Th alone produces heat $\sim 669.3 \times 54.555 = 36{,}514$ MeV/p, and in case of ^{233}U, heat is ~242,150 MeV/ p for the assumption that $k_{\text{eff}} = k = 0.982$. The weighted average of energy due to the mass-mixture of ^{232}Th and ^{233}U corresponding to the given thickness is expected to yield 40,444 MeV/p which is equivalent to energy gain, $G \sim 40.44$.

3.3.1 IAEA Benchmark Design

In order to access the basic requirements of an ADSS system, i.e., beam, target system, neutron fluence, and strategy of energy and transmutation of the SNF, IAEA floated [63, 64] a hypothetical design of 1,500 MW$_{\text{th}}$ ADS reactor (see Fig. 3.8) based on the Th-U fuel cycle. This is termed as IAEA benchmark. The IAEA-ADS benchmark provides an opportunity of mathematical modeling of a future ADS that may help in conducting various experiments and to settle down various questions which normally arise for the design and modeling of the device. The questions can be what will be the (i) neutron flux distribution in different parts/regions of the benchmark assembly (ii) neutron multiplication factor, (iii) transmutation potential (iv) criticality and (v) produced energy distribution. In this direction, a large number of simulation calculations

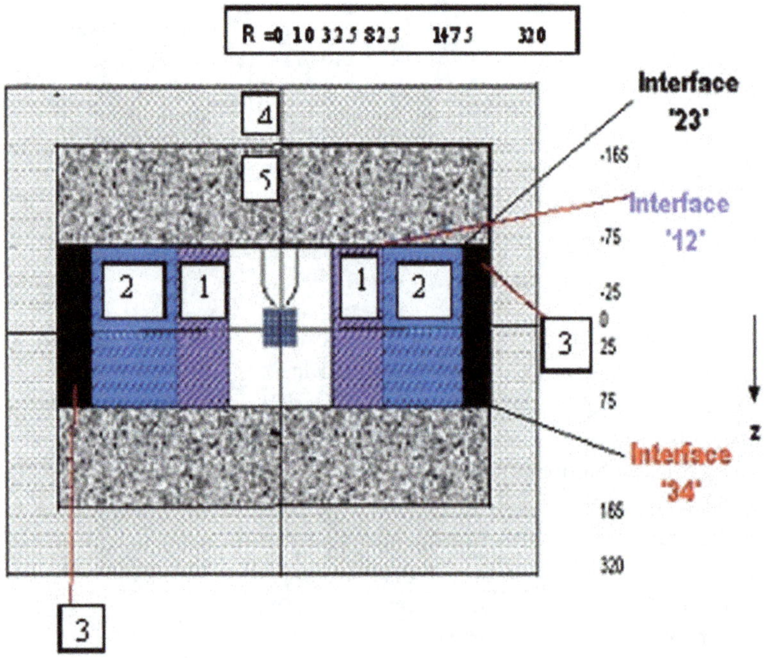

Fig. 3.8 XZ-cross-sectional view of the ADS benchmark design. The whole structure is in cylindrical shape of dimensions $2R \times L = 640 \times 640$ cm^2 [63, 64]

have been performed in different laboratories all over the world.

At first stage of the benchmark, verification of reactivity, burn-up swing as well as some reactivity effects for a fast neutron spectrum from an external spallation-type neutron source at different subcriticality levels like $k_{\mathrm{eff}} = 0.94$, 0.96, and 0.98 are proposed [63] for the development of a 1,500 $\mathrm{MW_{th}}$ ADS. In this direction, an important work for determination of concentration of ^{233}U with respect to ^{233}U + ^{232}Th fuel mixture is done by Carminati and Kady [63]. They estimated percentage of concentration of ^{233}U in ^{233}U + ^{232}Th mixture using JENDL-3.2 and JEFF-2.2 nuclear data libraries for the criticality $k_{\mathrm{eff}} = 0.982$, 0.962, and 0.941. Similarly, an important simulation using MCNP code [36] for neutron multiplication and flux distributions is done by Tueck et al. [64]. They also estimated the concentration of ^{233}U in the core region as function of k_{eff} using MCNP code. It is pointed out when the proportion of ^{233}U increases with respect to ^{232}Th, then k_{eff} improves and it moves toward the criticality level. They have also plotted ratio of spectrum averaged fission cross sections $\langle \sigma_f\,(^{232}\mathrm{Th})/\sigma_f\,(^{233}\mathrm{U})\rangle$ as function of radial position and established that for lower k_{eff}, the ratio is high at origin of the core and the ratio decreases toward outer region. Also, for simulation of transmutation and burn-up, an advance code Monte Carlo continuous energy burn-up code (MCB) developed by Cetnar [65] is used. The code is a combination of MCNP-4B code [36] for burn-up calculation and transmutation trajectory analysis (TTA) code [66] for the transmutation calculations. They have studied transmutation of ^{99}Tc and compared with the results of simulation from the MCNP-origin [36] computation system, and it is shown that the results are in good agreement. In another MC simulation approach using the Dubna CASCADE code [29, 35] alone, spallation n-distribution and power density as a function of radial distance and criticality [51] are calculated and results are found to be in good qualitative agreement with results presented by Tueck et al. [64] using the MCNP code. Some of the features of characteristic ADS benchmark reactor and results of calculations by Kumar [51] using the CASCADE code ver. 2004 are discussed in the following.

A cross-sectional view of hypothetical design of ADS benchmark reactor is shown in Fig. 3.8 with fuel and shielding features. At its center, a spallation lead target of dimensions $d \times L = 20$

Table 3.7 Nuclei densities (BOL at 20 °C) of elements in the five regions of the benchmark

Nuclei	Region 1	Region 2	Region 3	Region 4	Region 5
^{232}Th	–	–	7.45E−3	–	–
^{233}U + ^{232}Th	6.35E−3	7.45E−3	–	–	–
O	1.27E−2	1.49E−2	1.49E−2	–	–
Fe	8.10E−3	8.87E−3	8.87E−3	–	6.63E−3
Cr	1.12E−3	1.06E−3	1.06E−3	–	8.00E−4
Mn	4.60E−5	5.10E−5	5.10E−5	–	3.80E−5
W	4.60E−5	5.10E−5	5.10E−5	–	3.80E−5
Pb	1.77E−2	1.56E−2	1.56E−2	3.05E−2	2.41E−2

50 cm^2 is placed in vacuum spread over 32.5 cm radial distance. A beam pipe is shown just above the target. Regions marked as '1' and '2' are filled with different mixtures of ^{232}Th and ^{233}U fuels. Region '3' is filled with ^{232}Th alone as a fuel. The compositional details of the five regions are given in Table 3.7.

In Table 3.7, elements other than the nuclear fuel ^{232}Th and ^{233}U correspond to the RCC structure, plenum and other structural materials. Using the spallation neutron spectrum of Fig. 3.6 and the Dubna CASCADE code ver. 2004 MC simulation of the transport of *n*-spectrum through the regions '1' to '3' have been performed [51] for different proportions of ^{232}Th and ^{233}U in regions '1' and '2' and ^{232}Th alone filled in region '3' to obtain a desired value of k_{eff}. In Table 3.8, data of fuel composition corresponding to the three values of k_{eff} is given from which it is evident that with the fissile enrichment, k_{eff} increases.

3.3.2 Neutron Flux in Different Regions

In order to investigate the possibility of an appropriate region for the transmutation of a kind of a LLNW, what kind of neutron fluence is required. In a MC simulation by the CASCADE ver. 2004 code, a narrow detector is assumed at the interfaces of regions '1', '2', and '3' while building the input file of the code. An example of building such file is described in appendix of reference [47].

(i) **Target region**: In the target region, accelerated proton beam of energy 1 GeV collides with the lead target to produce spallation neutrons. Energy spectrum of neutrons is presented in Fig. 3.6. Axial distribution of produced neutrons is not uniform all along the *Z*-axis but skewed showing maxima at axial distance *Z* = 15–20 cm [47].

(ii) **Neutron distribution at interfaces of 1, 2, 3, and 4 regions**: Neutron distributions at the boundaries between the two regions are calculated assuming presence of a neutron detector of 1 cm width placed in between the two adjoining regions. In Fig. 3.9, neutron flux (n/cm^2/p) has been plotted for the '12', '23', and '34' boundaries for the three k_{eff} values by changing the fuel composition as given in Table 3.8. In the insets

Table 3.8 Different enrichment of ^{232}Th by ^{233}U and effective neutron multiplication, k_{eff}

k_{eff}	Region 1			Region 2		
	^{233}U	^{232}Th	^{233}U/[^{233}U + ^{232}Th] (%)	^{233}U	^{232}Th	^{233}U/[^{233}U + ^{232}Th] (%)
0.923	0.0113	0.1266	8.19	0.0127	0.1426	8.18
0.964	0.0121	0.1257	8.78	0.0137	0.1416	8.82
0.979	0.0125	0.1253	9.07	0.0141	0.1412	9.08

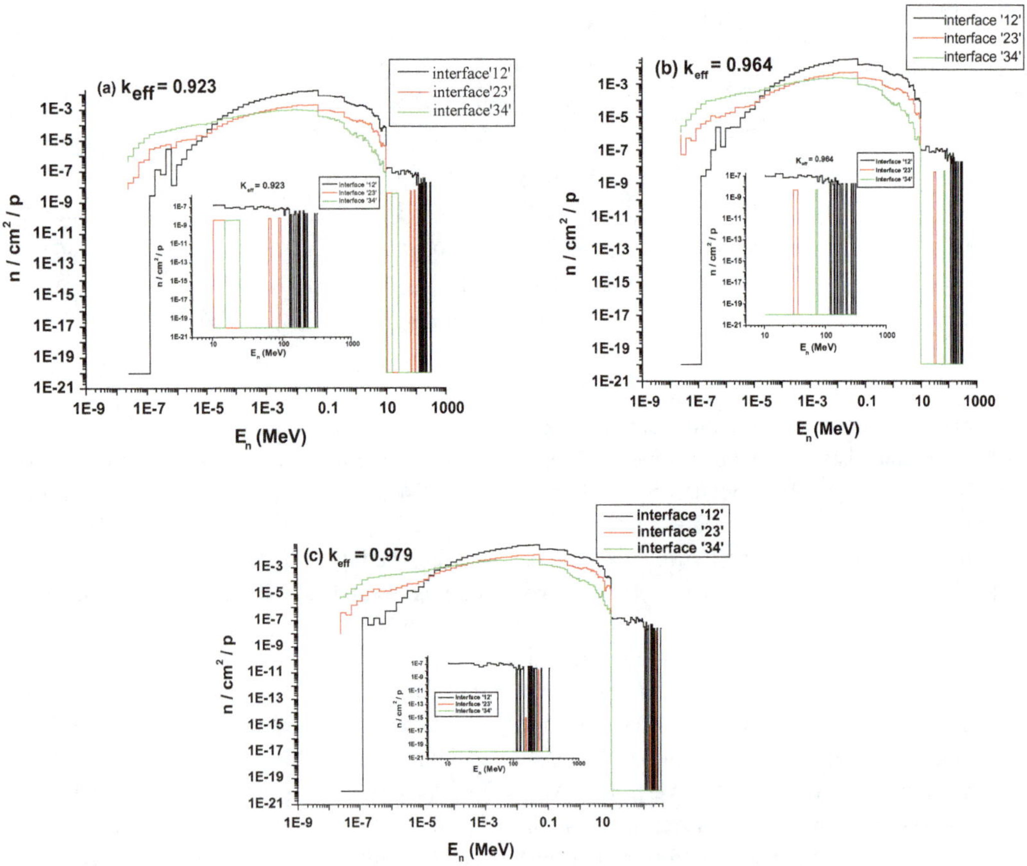

Fig. 3.9 Neutron fluxes (n/cm^2/p) at the three interfaces of the IAEA benchmark with k_{eff} = 0.923, 0.964, and 0.979 calculated from the CASCADE ver. 2004 are displayed [47]

of the three (a), (b) and (c), plots elaborated distributions of high-energy neutrons from 10 MeV to highest energy are given.

For summarizing the neutron growth at boundaries, '12', '23', and '34', the flux is categorized in the following three ranges and listed in Table 3.9. The flux corresponds to multiplication of spallation neutrons of 1 GeV p + Pb collision after transport through the fuel and structure materials assumed in the IAEA benchmark design as given above.

(i) Range I: E_n = 0.25 eV < $E_n \leq$ 1 eV (thermal + epithermal range)
(ii) Range II: 1 eV < $E_n \leq$ 0.1 MeV (resonance range) and
(iii) Range III: 0.1 MeV < $E_n \leq$ maximum energy

From the flux data displayed in Table 3.9, the following observations can be made:

(a) For the three k_{eff} values, flux of thermal + epithermal neutrons of range I increases from inner region to the outer region, while neutron fluxes at other two higher energies decrease.
(b) Total neutron flux reaching boundary '34' from the boundary '12' increases as 6.3, 8.5, and 9.4% for the three cases of k_{eff} = 0.923, 0.964, and 0.979, respectively. Naturally, always more than 6% neutrons are reaching to the shielding wall, i.e., region 4 of the proposed benchmark design.

Table 3.9 Number of neutrons per unit area per incident proton (n/cm²/p) and absolute flux (n/cm²/s) at interfaces '12', '23', and '34' for k_{eff} = 0.923, 0.964, and 0.979 in the three neutron energy ranges

k_{eff}, beam current		n/cm²/p in neutron energy range			Total (n/cm²/p)	Flux (n/cm²/s)
		$0.025 < E_n \le 1$ eV	1 eV $< E_n \le 0.1$ MeV	$0.1 < E_n \le 326$ MeV		
0.923, 27.8 mA	'12'	3.20E−06	1.51E−01	0.97E−01	0.248	4.3152E+16
	'23'	2.21E−05	1.99E−02	9.92E−03	0.0298	5.1852E+15
	'34'	2.22E−04	1.33E−02	2.27E−03	0.0158	2.7492E+15
0.964, 10.93 mA	'12'	2.55E−06	2.89E−01	1.95E−01	0.484	3.3057E+16
	'23'	4.37E−05	5.07E−02	26.0E−03	0.0767	5.2386E+15
	'34'	5.70E−04	3.42E−02	6.15E−03	0.0409	2.7934E+15
0.979, 5.98 mA	'12'	9.26E−07	4.51E−01	3.08E−01	0.759	2.8386E+16
	'23'	8.41E−05	8.75E−02	45.90E−03	0.134	5.0116E+15
	'34'	1.07E−03	5.85E−02	11.0E−03	0.071	2.6554E+15

(c) In case of $k_{eff} = 0.979$ criticality, among the total neutron spectrum reaching at the boundary '34', there are ~15.5% neutrons with high energy $0.1 < E_n \le 326$ MeV. They can be highly useful for transmutation of elements of the nuclear waste. Already, in the structure, thorium is filled in region '3' to be burnt by the hard spectrum. At the same time, it helps in stopping the outgoing flux.

(d) At $k_{eff} = 0.979$, neutron flux escaping to the shielding through the boundary '34' is unique and nearly flatly distributed from the thermal to the fast reactor energies. Also, it is ~30 times more than the flux of a thermal reactor. In fact, this enhances possibility that a strong thermal reactor can be allowed to operate on the basis of this neutron strength.

3.3.3 Accelerator Current and Radial Heat Distribution

One ampere current is equivalent to 6.25×10^{18} p/s and for the assumed heat production of 54,000 MeV/p, there will be heat rate 3.375×10^{23} MeV/s which is equivalent to 5.4×10^4 MW. For a hypothetical reactor of 1,500 MW_{th}, one needs 27.8 mA current and this corresponds to $k = 0.923$. It will be much less current ~10.93 mA for $k = 0.964$ and 5.98 mA in case of $k = 0.979$ respectively for the same power of the reactor. Corresponding to the three current values, the proton beam intensities are 1.74×10^{17} (p/s), 6.83×10^{16} (p/s), and 3.74×10^{16} (p/s), respectively. Neutron flux calculated for the 27.8, 10.93, and 5.98 mA currents for the three criticalities k_{eff} of the reactor are also given in the last column of Table 3.9 separately for the three boundaries between four regions of the reactor shown in Fig. 3.8. Radial heat distribution is also calculated using the CASCADE code ver. 2004 code [51] and given in Fig. 3.10 showing a peak in the

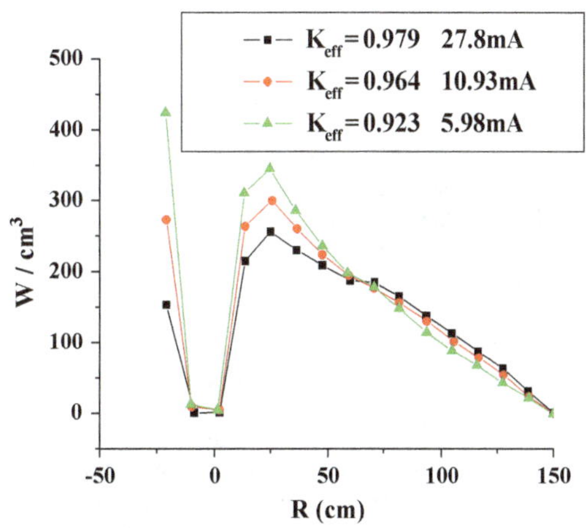

Fig. 3.10 Heat density (W/cm³) as a function of radial distance R of the IAEA benchmark design for the three criticalities k_{eff} [51]

target region and another peak at $R \sim 50$ cm in the fuel region as the fuel ranges from $R = 32$ to 150 cm. Heat distribution corresponding to $k_{eff} \sim 0.979$ is flatter than at smaller k_{eff} and varies from 0 to 350 W/cm^3.

In the end, it may be concluded that the calculated neutron flux in different positions of a reactor can be exploited for the investigations of transmutation of the LLNW and utilization of fertile fuels like ^{232}Th and ^{238}U.

References

1. Bothe, W., Becker, H.: Künstliche erregung von kern-γ-strahlen. Z. Physik. **66**, 289 (1930)
2. Chadwick, J.: Proc. Royal Soc. (London). Ser. A **136**, 692 (1932). Lorch, E.A.: Original Am-Be neutron source spectrum. Int. Jpn. Appl. Radiat. Isot. **24**, 585 (1973)
3. Serber, R.: Nuclear reactions at high energies. Phys. Rev. **72**, 1114 (1947)
4. Rubbia, C., et al.: Conceptual design of a fast neutron operated high power energy amplifier. CERN Report CERN/AT/95-44 (ET)
5. Rubbia, C.: A high gain energy amplifier operated with fast neutrons. In: AIP Conference Proceedings 346, International Conference on Accelerator-Driven Transmutation Technologies and Applications, Las Vegas (NV), US (1994)
6. SINQ: The Swiss Spallation Neutron Source. http://www.psi.ch/sinq/
7. ISIS: ISIS Neutron and Muon Source. http://www.isis.stfc.ac.uk
8. SNS: Neutrons for New Discoveries and Solutions. http://neutrons.ornl.gov
9. Futakawa, M., et al.: 1-MW Pulsed Spallation Neutron Source (JSNS) at J-PARC. Neutron News **22**, 15–19. JSNS. http://j-parc.jp/MatLife/en/index.html (2011)
10. Henderson, S., et al.: The Spallation Neutron Source accelerator system design. Nucl. Inst. Methods Phys. Res. A. **763**, 610 (2014)
11. David, J.-C.: Spallation Reactions. A Successful Interplay between Modeling and Applications. arXiv: 1505.03282v1 [nucl-ex], 13 May 2015
12. Kumar, V., eta al.: Neutron spallation source and the Dubna Cascade Code. Pramana-Jpn. Phys. **60**, 469 (2003)
13. Hilsher, D., et al.: Neutron multiplicity distribution for GeV proton induced spallation reactions on thin and thick targets of Pb and U. In: Proceedings of the International Workshop on Nuclear Methods for Transmutation of Nuclear Waste, edited by Khankhasayev, M.Kh., Plendl, H.S., Kurmanov, Z.B., p. 176 (1996)
14. Lee, Y.Y., et al.: Proc. ICANS-XIII PSI-Proc. **95** (02), 802 (1995)
15. Polański A., Słowiński, B.: Neutron fields produced in heavy spallation targets by electron beams. In: International Conference on Recent Developments and Applications of Nuclear Technologies, 15–17 September 2008. Also see, Polański, A., Słowiński, B., Jackowski, T., Pacan, A.: Neutrons production in heavy extended targets by electrons of energy from 15 to 1000 MeV. Prog. Nucl. Energy **78**, 1 (2015). Also see http://dx.doi.org/10.1016/j.pnucene.2014.07.009
16. Auditore, L., et al.: Study of a 5 MeV electron linac based neutron source. Nucl. Instr. Methods Phys. Res. B **229**, 137 (2005)
17. Roberts, J.A.: The Los Alamos Neutron Science Center (LANSCE). Neutron News **10**(4), 11 (1999)
18. Mason, T.E., et al.: Spallation Neutron Source in Oak Ridge: a powerful tool for materials research. Phys. B **385**, 955 (2006). Also see, http://neutrons.ornl.gov/
19. Gabriel, T., et al.: Overview of the Spallation Neutron Source (SNS) with emphasis on target systems. J. Nucl. Mater. **318**, 1 (2003)
20. Eccleston, R., Wilson, C.: Correspondent's reports: a guided tour of ISIS: 2004 update. Neutron News **15** (1), 15 (2004)
21. Penfold, J.: Scientific review: ISIS second target station: overview and progress. Neutron News **15**(2), 9 (2004)
22. Wagner, W., et al.: The Swiss spallation neutron source SINQ—developments and upgrades for optimized user service. Phys. B **385**, 968 (2006)
23. Wagner, W.: The Swiss Spallation Neutron Source SINQ-Layout, Operation, Utilization and R&D for Optimized Neutron Yield. Proc. ICANS-XVI **53** (2003). ISSN 1433-559X
24. Arai, M., Maekawa, F.: Japan Spallation Neutron Source (JSNS) of J-PARC. Nucl. Phys. News **19**(4) (2009)
25. Maekawa, F., et al.: First neutron production utilizing J-PARC pulsed spallation neutron source JSNS and neutronic performance demonstrated. Nucl. Inst. Methods Phys. Res. A **620**, 159 (2010)
26. Westmeier, W., et al.: Transmutation experiments on ^{129}I, ^{139}La and ^{237}Np using the Nuclotron accelerator. Radiochim. Acta **93**, 65 (2005)
27. Adam, J., Barashenkov, V.S., et al.: Measurement of the neutron fluence on the spallation source at Dubna. Kerntechnik **70**, 127 (2005)
28. Krivopustov, M.I., et al.: First experiments on transmutation studies of ^{129}I and ^{237}Np using relativistic protons of 3.7 GeV. J. Radioanalytical Nucl. Chem. **222**, 267 (1997). JINR Dubna preprint E1-97-59. Krivopustov, M.I., et al.: First experiments with a large uranium blanket within the installation energy plus transmutation exposed to 1.5 GeV protons. Kerntechnik **68**, 48 (2003)
29. Barashenkov, V.S., et al.: Mathematical model of the electronuclear set-up on the beam of JINR synchrotron. Nucl. Inst. Methods B **217**, 352 (2004)

30. Adam, J., et al.: A study of nuclear transmutation in Pb-target and U-blanket irradiated by 1.6 GeV deuteron. Eur. Phys. J. A **43**, 159 (2010). Adam, J. et al.: A study of reaction rates of (n, f), (n, g) and (n, 2n) reactions in U^{nat} and Th^{232} by the neutron fluence produced in the graphite set up (GAMMA 3) irradiated by 2.33 GeV deuteron beam. Eur. Phy. J. A **47**, 85 (2011)
31. Adam, J., et al.: Measurement of the high energy neutron flux on the surface of the natural uranium target assembly QUINTA irradiated by deuterons of 4 and 8 GeV energy. Phys. Proc. **80**, 94 (2015)
32. Shvetsov, V., et al.: The Subcritical Assembly in Dubna (SAD)-Part I: coupling all major components of an Accelerator Driven System (ADS). Nucl. Instr. Methods Phys. Res. A **562**, 883 (2006). Gudowski, W., et al.: The Subcritical Assembly in Dubna (SAD)-Part II: research program for ADS-demo experiment. Nucl. Instr. Methods Phys. Res. A **562**, 887 (2006)
33. Abanades, A., ,et al.: Results from the TARC experiment: spallation neutron phenomenology in lead and neutron-driven nuclear transmutation by adiabatic resonance crossing. Nucl. Instr. Methods Phys. Res. A **478**, 577 (2002)
34. Borcea, C., et al.: Results from the commissioning of the n_TOF spallation neutron source at CERN. Nucl. Instr. Methods Phys. Res. A **513**, 524 (2003)
35. Barashenkov, V.S.: Monte Carlo simulation of ionization and nuclear processes initiated by hadron and ion beams in media. Comp. Phys. Commun. **126**, 28 (2000)
36. Breismeister, J.F. (ed.): LA-12625-M, MCNP (A General Monte Carlo N-Particle Transport Code, Version 4A). Los Alamos National Laboratory, Los Alamos, New Mexico, issued (November, 1993)
37. Tagliente, G.: n_TOF collaboration: the n_TOF facility at CERN. Brazilian J. Phys. **34**(3A) (September, 2004)
38. Maglioni, C.: CERN EN-STI-TCD, Pb/Bi loop target for EURISOL kick-off meeting, 5/10/2012. http://pceet075.cern.ch/ and http://www.efnudat.org/FZD_workshop/Wednesday/1302-1000_Vlachoudis.pdf
39. Neutron Cross Section Measurements at n-TOF for ADS Related Studies. http://iopscience.iop.org/article/10.1088/1742-6596/41/1/038/pdf;jsessionid=4A1B8122075DBA37D070F7B8C2B513DE.c5.iopscience.cld.iop.org
40. Snow, W.M.: Neutron Physics. https://www.bnl.gov/npss/files/pdf/SnowNPSS1.pdf
41. Nelson, R.O., et al.: Stockpile stewardship and nuclear science with GEANIE at LANSCE/WNR, LLNL. Physics Division, Progress Report, Chapter 2, 64. http://wnr.lanl.gov/; http://www.geanie.lanl.gov/ (1996)
42. Bol, A., et al.: A novel design for a fast intense neutron beam. Nucl. Instr. Meth. **214**, 169 (1983)
43. Kerveno, M., et al.: Measurement of ^{232}Th (n, 5nγ) cross sections from 29 to 42 MeV. In: International Conference on Nuclear Data for Science and Technology (2007). https://doi.org/10.1051/ndata:07415
44. Kim, G.N.: Proceedings of LINAC, p. 383, Gyeongju, Korea. http://accelconf.web.cern.ch/AccelConf/l02/PAPERS/TU432.PDF (2002)
45. Salomé, J.M., et al.: Neutron producing targets at GELINA. Nucl. Instr. Methods **179**, 13 (1981). Mihailescu, L.C., et al.: A new HPGe setup at Gelina for measurement of gamma-ray production cross-sections from inelastic neutron scattering. Nucl. Instr. Methods Phys. Res. A **531**, 375 (2004)
46. Antropov, V., et al.: IREN Test Facility at JINR. http://linac96.web.cern.ch/Linac96/Proceedings/Tuesday/TUP68/Paper.html
47. Sharma, M.: Transmutation of long-lived isotopes of conventional reactors using A.D.S. Concept. Ph.D. thesis, University of Rajasthan, Jaipur (2010)
48. Katovsky, K.: Investigation of secondary neutrons and nuclei generated in reactions of high-energy protons and neutrons in uranium and plutonium targets. Ph.D. thesis, CTU, Prague (2008)
49. Bhatia, C., Kumar, V.: Role of (n, xn) reactions in Accelerator Driven Sub critical Systems. Published by LAP LAMBERT Academic Publishing GmbH & Co. KG (2011)
50. Kumar, V., et al.: CASCADE data for the A.D.S. materials for it benchmarking. In: Proceedings ARIA 2008, 1st Workshop on Accelerator Radiation Induced Activation, 13–17 October 2008, Paul Scherrer Institute, Switzerland. PSI Proceedings 09–01. ISSN 1019–0643, 30 (2009)
51. Kumar, V.: Role of (n, xn) reactions in ADS, IAEA-Benchmark and the Dubna CASCADE Code Pramana. J. Phys. **68**, 315 (2007)
52. Koning, A.J., et al.: Proceedings of the International Conference on Nuclear Data for Science and Technology, Santa Fe, USA, 26 September–1 October 2004 and www.talys.eu/
53. Barashenkov, V.S., et al.: CASCADE program complex for Monte-Carlo simulation of nuclear processes initiated by high energy particles and nuclei in gaseous and condensed matter. JINR preprint R2-85-173 (1985). Barashenkov, V.S.: A cascade-evaporation model for photonuclear reactions. Nucl. Phys. A **231**, 462 (1974)
54. Kim, E., et al.: Measurements of neutron spallation cross sections of ^{12}C and ^{209}Bi in the 20- to 150-MeV energy range. J. Nucl. Sci. Eng. **129**, 209 (1998)
55. Kim, E., et al.: Measurements of Neutron Spallation Cross Sections of ^{12}C and ^{209}Bi in the 20- to 150-MeV Energy Range. J. Nucl. Sci. Engg. **129**, 209 (1998)
56. Uddin, M.S., et al.: Measurements of neutron induced activation of concrete at 64.5 MeV. Ann. Nucl. Energy **36**, 1133 (2009)
57. Accelerator Driven Systems (ADS) and Fast Reactors (FR) in Advanced Nuclear Fuel Cycles.

A Comparative Study. www.nea.fr/html/ndd/reports/2002/nea3109.html
58. See Exfore entry 33004001 on site http://www-nds.indcentre.org.in/EXFOR/33005.003
59. Cullen, D.E., et al.: ENDF/B-VII.0 Data testing for three fast critical assemblies. Report UCRL-TR-233310 (2007). Cullen, D.E.: PREPRO 2000:2000 ENDF/B pre-processing codes. Report IAEA-NDS-39, Rev. 10, 1 April 2000
60. Bhatia, C., Kumar, V.: Determination of neutron multiplication coefficient of the fuel elements irradiated by the spallation neutrons. Phys. Rev. C **81**, 024614 (2010)
61. Dudnikov, A., Sedov, A.A.: RBEC-M lead-bismuth cooled fast reactor benchmarking calculations. International Atomic Energy Agency. http://www.iaea.org/OurWork/ST/NE/NENP/NPTDS/Downloads/SMR_CRP1_SRWOSR/2006/RBEC-M%20Kurchatov%20Final.pdf\
62. Thermal Neutron Capture Cross sections and resonance integrals—Fission product nuclear data. http://www-nds.ipen.br/sgnucdat/b3.pdf
63. Carminati, F., Kadi, Y.: ADS Neutronic Benchmark (stage 1): a new approach to the design of accelerator driven systems. In: Proceedings of the International Atomic Energy Agency Technical Committee Meeting, Madrid, Spain, 17–19 September 1997
64. Tucek, K., et al.: IAEA Accelerator Driven System Neutronic Benchmark. In: Proceedings of the International Atomic Energy Agency Technical Committee Meeting, Madrid, Spain, 17–19 September 1997
65. Cetnar, J.: A method of transmutation trajectories analysis in accelerator driven system. In: Proceedings of IAEA Technical Committee meeting on Feasibility and Motivation for Hybrid Concepts for Nuclear Energy Generation and Transmutation, Madrid, 17–19 September 1997
66. Cox, L.J.: LA-UR-11-01654: LNK3DNT Geometry Support: User Guidance for Creating and Embedding and https://rsicc.ornl.gov/

4 Requirement of Nuclear Data

In this chapter, we discuss requirement of nuclear data from the point of accuracy and sectorial demand of different kinds of data for the development of the ADSS technology. While writing the chapter, we realized that compilation of all kinds of data including evaluated data and parameterization of data is a matter of another book. Therefore, advancement of new data built up and related developments have been discussed in the chapter. A few data of simulated displacement cross sections of radiation damage and gas production in the target section of the ADSS are presented in Sect. 4.2.2 of the chapter. These data need their experimental validation.

4.1 Requirement of Nuclear Data Beyond the Existing Power Reactors

As an initial step, electronuclear systems had been extensively discussed first at Lawrence Livermore Laboratory in the year 1976 [1] under the material test accelerator (MTA) project for producing fissile fuels like ^{239}Pu and ^{233}U from the corresponding fertile fuels by making use of spallation neutrons from the deuteron beams of energy 500–600 MeV. Similarly, Canadian, intense neutron generator (ING) [2] project aimed at breeding of fissile fuel from the fertile fuel using the accelerated particle beam was initiated. Depending on the initial calculations and a few experiments conducted on accelerators, a long list of research and development programs has been made at the national and international levels. A technologist can realize the amount and the kind of data requirement seeing at the schematic diagram of accelerator-driven accelerator system shown in Fig. 4.1 [3].

These programs have been discussed in different chapters of the book. The programs direct more or less on the feasibility studies of measurement of neutron flux and have shown concern of the spent fuel world over. The new technology of ADSS is many times more complicated than designing of a power reactor in the last century. Ikeda [4] has identified that the nuclear data related to hundreds of elements pertaining to target material, fuel elements, fission products, fuel cells, structural materials, and shielding, etc., need to be made available at much higher neutron energies and with much better precision of cross section measurements for many physics aspects of the design. Chemical and metallurgical aspects have other dimensions of the technology.

According to Plompen [5], 'while discussing about the need of nuclear data again raises several issues of quantitative precision of the data and clever high-quality calculation/simulation.' Among the nuclear challenges, the following are counted:

(i) fissile nuclide capture cross section needs to be measured with <5% accuracy

V. K. Verma and K. Katovsky, *Spent Nuclear Fuel and Accelerator-Driven Subcritical Systems*, Green Energy and Technology, https://doi.org/10.1007/978-981-10-7503-2_4

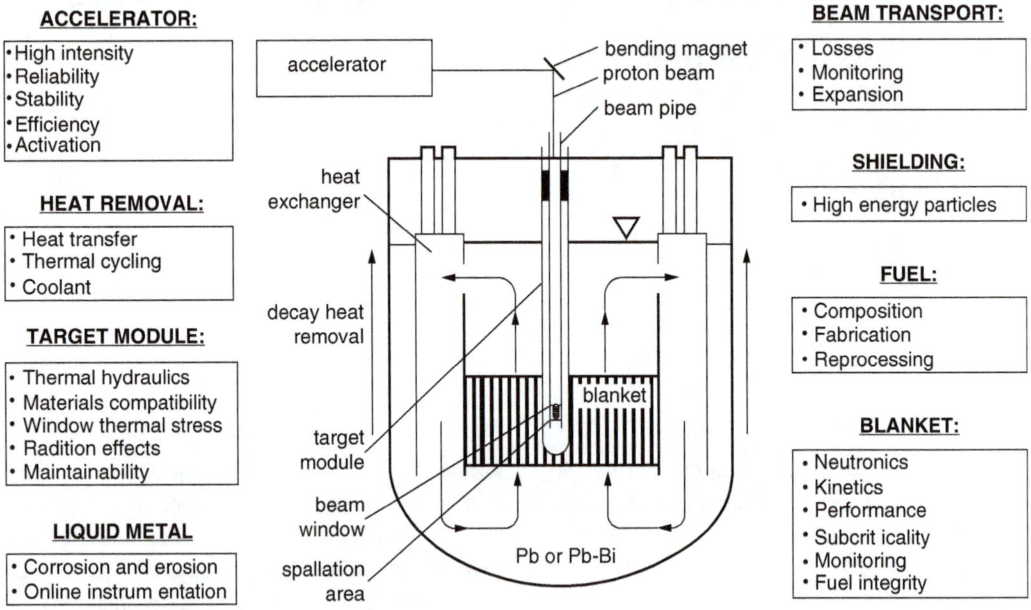

Fig. 4.1 A schematic diagram of ADSS with special components [3]

(ii) scattering cross sections to be measured with 2–5% and fission cross sections with ~2% accuracy in case of minor actinides (MA)

(iii) number of fission neutrons to be measured with accuracy and

(iv) overall neutron spectrum to be measured and estimated.

Effect of accuracy of the data can be understood effectively after an elaborate discussion of new reaction channels opened up by the high-energy neutrons because accuracy of energy efficiency, ΔG of the ADSS depends strongly on the accuracy of cross section measurements as can be seen from the following discussion. Energy efficiency of the ADSS is related to the energy gain, G, and this is a function of neutron multiplication factor, k_s.

$$G = G_0\left(\frac{k_s}{1-k_s}\right) \text{ with } G_0 = \left(\frac{\langle E_f\rangle \times n_s}{E_p \times v}\right) \tag{4.1}$$

where $\langle E_f\rangle$ = fission energy, E_p = proton beam energy, n_s = number of spallation neutrons, v = fission neutrons. Neutron multiplication factor k_s is defined by Eq. (3.16) is equal to production term, P. In case, there is a leak-free system where all neutrons are utilized, then $R + L = 1$ and $k_s = P$. Thus, the accuracy of production term, P, defined as Eq. (3.14) is responsible for the accuracy of gain, G. Equation (3.16) can be rewritten as follows:

$$P = [1 - C]\frac{A}{B} \tag{4.2}$$

where

$$A = \left[2\Sigma_{n,2n} + 3\Sigma_{n,3n} + \cdots + \langle \upsilon\rangle\Sigma_{n,f} + 3\Sigma_{n,3n\alpha} + \cdots\right]$$

$$B = \Sigma_t \quad \text{and} \quad C = \mathrm{e}^{-x\Sigma_t}$$

Symbol Σ corresponds to the macroscopic cross section of a reaction shown as the suffix of the symbol, and $\langle v\rangle$ is the average number of fission neutrons.

This leads to

$$\Delta P^2 = \frac{A^2}{B^2}\Delta C^2 + (1 - C)^2\left\{\frac{\Delta A^2}{B^2} + \frac{A^2}{B^4}\Delta B^2\right\} \tag{4.3}$$

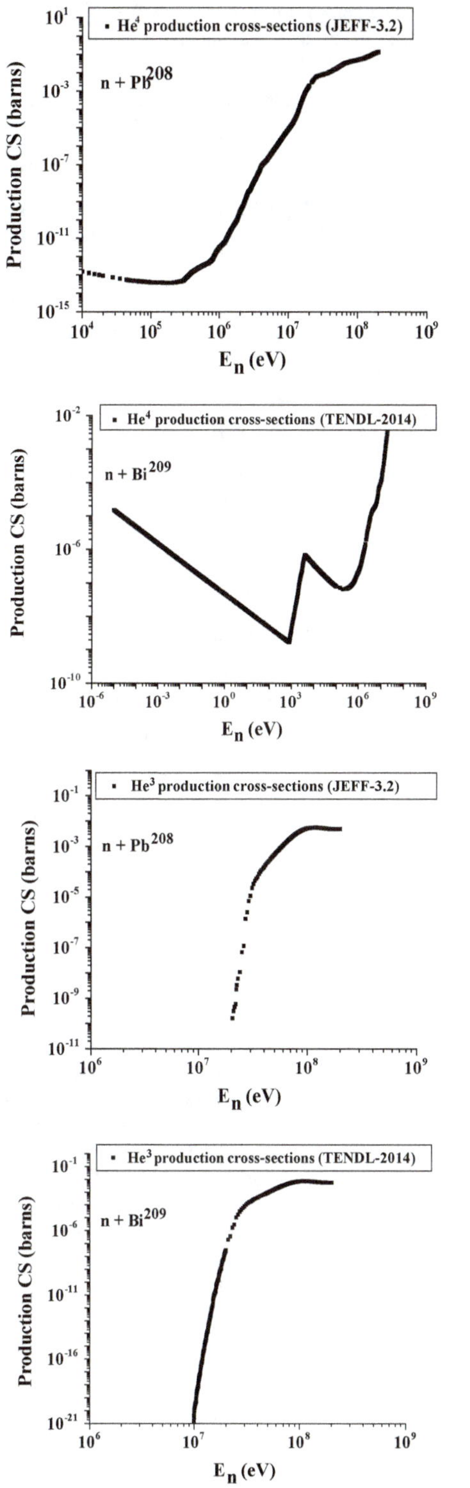

Fig. 4.2 He^4 and He^3 gas production cross sections in *n* + Pb and *n* + Bi collision plotted as function of neutron energy [34–36]

$$\Delta C = \pm \Sigma_t e^{-x\Sigma_t} \Delta x \tag{4.4}$$

Thus, the accuracy of determination of production term, *P*, shows the following behavior:

(i) Increase of term *B* or increase of the total cross section, Σ_t is helpful in increasing accuracy of the term, *P*.
(ii) Increase of the term *B* corresponds to decrease of ΔC due to exponential part of the term will be helpful automatically.
(iii) Decrease of the terms *A*, ΔA, ΔB, and ΔC will be helpful in increasing accuracy.

Better accuracy means, ΔP is smaller; hence, the energy gain, ΔG, can be estimated precisely. This automatically emphasizes need of better experiments by using fast techniques of measurements, new analytical methods along with need of regular check of data for individual need and to transfer to the IAEA, IAEA-NDS and similar other Web sites.

4.2 Nuclear Data

Fast reactors and ADSS are different in many ways, and the major difference arises due to neutron energy and the target system which is a source of neutrons. Some technical part of the target system is discussed elsewhere in the book. Both fast neutrons and spallation neutron spectra have peaks around 2–4 MeV, but the spallation neutron spectrum is extended up to the projectile beam energy compared to fast reactor neutron energy which exists at ~10.5 MeV. Also, according to the CASCADE code calculations, projected in Fig. 3.6 of the last chapter, spallation neutrons produced by 1 GeV proton colliding with the lead target are ~8.68% having energy ≥10.5 MeV. They multiply in the fuel ahead to them by way of (*n*, *xn*) kind of reactions besides the fission process. In Table 4.1, an attempt is made to identify and summarize large amount of the data required for the ADSS technology.

Table 4.1 Requirement of data of different kinds of target, container, *n*-poisons, fuel cycles, shielding, and radiation-resistant materials with required precision level as discussed in the text. CS stands for cross section. New requirement of measurements is shown in boldface

System sector	Material elements	Elements (at. mass)	Specific data requirement and references
Target system			
Solid	Pb, Bi, W, Ta, Hg, ^{nat}U, Th	Pb (202, 204, 206, 207, 208); Bi (209); W (182, 183, 184, 186); Ta (181), Th, ^{nat}U	CS {(*p*, reac), (*n*, *n*′), (*n*, *xn*), (*n*, *xp*), (*p*, *xp*), (*p*, *f*), (*n*, *f*)} with required accuracy [6]
Eutectic	Pb + Bi (ppm level of Ag, Fe, Ni, Sn, Cd, Al, Cu, and Zn)		Heat conduction, neutron poison, study of corrosion and erosion etc.
Container	Al, Sn, Fe, Cr, Ni, Mo, Zr, V, Ta, W, Cl, Na etc.	Al (27); Sn (112,114–120, 122, 124); Fe (54, 56, 57, 58); Zr (90, 91, 92, 94, 96); Cl (35, 37); Na (23)	**1. More CS data on (n, α) and (n, p) and other gas production by HE n's and protons and** **2. Threshold displacement energy, E_d**
n-poison and toxic gases like Hg and He^4, He^3 and H^1	Po and other *n*-poisons	Po (207–209) in Bi (*p*, *xn*) Po (210) in Bi (*n*, *γ*)	1. CS {(*p*, *xn*), (*n*, *γ*), (*p*, *xp*)} [6] 2. Po is not a poison for HE neutrons of ADSS 3. Search of new *n*-poisons for HE spallation neutron spectrum 4. Release of gases, He^4, He^3, and H^1 in target system and 5. Presence of Po and Hg in cover gas
Structural material			
Fuel cells	Al, Mg, Si, Fe, Cu, Ti, Zn, Cr, Zr, Si, Py, C, O		**1. Gas production data at fast to spallation neutron energies** **2. Radiation damage cross sections. and** **3. Threshold displacement energy, E_d**
Shielding	B, Be, Si, P, Cr, Fe, V, Mn, Ni, Zr, Mo, Nb, Sn, Zn, Cu, C, O, Ti, Ca, Ar, Mg, Na, O, Pb		**1. Corrosion, erosion, swelling, brittleness, gas production data, heat conduction** **2. Damage cross sections by both gamma and fast neutron spectra. Radiation resistance character of material / elements and** **3. Threshold displacement energy, E_d of several elements**
Gases	H, He, O		Gas production, swelling, reduction and oxidation rates are required
Beam window	T91/F91, HT9, EP82, 9Cr-1Mo steels	For example, T91/F91 steel has composition: Ni, Cr, Mn, Mo, Si, Ti, V, C, P, S, Nb, N, Al, Cu, As, Sn and Fe	Window materials have been discussed in Sect. 4.1.3.4 including graphene as recent one

(continued)

Table 4.1 (continued)

System sector	Material elements	Elements (at. mass)	Specific data requirement and references
Fuel and produced material			
Fuels	U-Pu, Th-U, N, O, F, Cl	U (234–237, 238), Pu (238–42), Am (241); ^{14,15}N, O, Cl and F	**1. (n, f), (n, xnyp), (n, γ) for HE spallation neutron spectrum in case of actinides Complete fuel cycle as discussed in Chap.** 5 **and** **2. Data of He, H and other gas production in (n, α) and (n, p) reactions by HE neutrons including the container material**
Produced actinides	Np, Am, Cm, Pu, U	^{237,238}Np, $^{241-243}$Am, $^{242-48}$Cm, $^{238-40}$Pu and ^{233}U	Production cross sections with required accuracy of spallation neutron spectrum are required
Long-lived fission products	Se, Zr, Tc, Pd, Sn, I, Cs	^{79}Se, ^{93}Zr, ^{99}Tc, ^{107}Pd, ^{126}Sn, ^{129}I, ^{135}Cs	1. Fission yield 2. Production CS of fission products **3. CS {(n, γ) and (n, xnyp)} of LLFP**

In the time of design of fast power reactors, emphasis on the measurement of interaction cross sections or reaction rates of high-energy neutrons was stressed and the issue of multiplication of neutrons beyond the $(n, 2n)$ reactions was not emphasized. Also, issues related to shielding were not so different to the thermal reactors. In case of ADSS, extension of fuel cycles, design of shielding, design of beam window, overall radio-toxicity and radiation damage are dimensionally different let alone the issue of design of the target system.

In this direction, there is spurt in development of Monte Carlo codes for computation of cross sections of high-energy neutrons using theoretical models, multiplication of neutrons, energy applications, and questions related to the damage of materials. In Chap. 6, development of experimental facilities world over has been discussed in this book where kind of data production has been main issue along with validation of the data by the experiments.

Requirement of data is divided in different sections in Table 4.1 as per material requirement of ADSS. Accuracy of data can be stressed upon in case of the utilization of existing old data and for the purpose of performing new experiments.

Several kinds of fast reactors have been designed, and they are running successfully. Development of generation IV reactors is also on the way, and both categories are utilizing fast reactor spectra. Both energy amplifier and a device of incineration of long-lived highly toxic reactor products, there is need of high energy neutron source. Understandably, in this environment, the fuel cycles of the subcritical reactors are going to be very much different to a normal critical reactor. At the same time, there will be need of much high shielding of the left-over or escaping out neutron flux. Enhancing energy efficiency of the subcritical system means increasing P or complete utilization of a neutron available in a system. This can be done when a neutron loses very small amount of energy by way of radiation damage and development of radiation-resistant materials will become a necessity. Requirement of data for all the major sectors of an ADSS design is discussed as follows.

4.2.1 Data Requirement of a Target System

The target system comprises of a solid or liquid target metal of high Z and high N which can generate high flux of neutrons and transmit them to the fuel blanket. Coupling of the beam through the beam window is also a part of the target system. In case a target is in liquid form like a eutectic, then it may be preferred that it can work as a coolant also. The entire target system is discussed from the data point of view with little focus on the other technical details such as kinetics of eutectic, architecture of the eutectic container, and similar other issues.

4.2.1.1 Target Cum Coolant

The target system is preferably a high Z, high N heavy metal like Pb, Bi, Hg, and even a fertile fuel in the metal phase. The target itself is enclosed in a metallic container. If the target is an eutectic flowing within a specific steel container and the whole target unit is irradiated by a high current (10–30 mA) of high-energy particle beam such as proton or deuteron or even electron. Its initial face toward the beam pipe will be irradiated heavily by the charged particle beam and exposed to heat, produced neutrons and other heavier particles. The nuclear processes responsible for neutron production in particle + A collision are pre-equilibrium, compound evaporation, direct as well as cascade emissions followed by the de-excitation of target nucleus. Irradiation by little proportion of nuclear fragments as well as fission fragments compared to earlier two cannot be denied. Many models and even big codes of particle production involving various cascade processes have been developed, and they have been extensively discussed in Refs. [7–9]. One of the commonly used MC codes, LAHET [10] has been used in several codes of particle transport and production of particles. According to the code, in case of 1 GeV proton colliding with Pb block of dimension say, $d \times L = 8 \times 50$ cm^2, total 22.3 neutrons are produced. Out of them, there are 16 neutrons due to evaporation and fission processes, and on an average, 6.3 neutrons are due to cascade process. Neutrons so produced from the target will multiply in the medium, and various nuclear processes are mentioned in Sect. 3.2. It may be mentioned that all these nuclear processes need updated data of cross section measurements for an accurate design of the ADSS system.

According to Glasbrenner et al. [11], LBE (44.8 wt% Pb + 55.2 wt% Bi) contains impurities of few ppm of elements like Ag, Fe, Ni, Sn, Cd, Al, Cu, and Zn. Along with a big list of data of various reactions with the elements of the eutectic, there is requirement of data of these impurity elements also. In irradiation of Pb–Bi by HE protons, production CS of many isotopes like ^{177}Lu, ^{183}Re, ^{183}Ta, ^{185}Os, ^{188}Ir, ^{188}Pt, ^{189}Ir, ^{191}Pt, ^{195}Hg, ^{195}Pt, ^{196}Au, ^{198}Au, ^{199}Au, ^{201}Tl, ^{202}Tl, ^{203}Hg, ^{203}Pb, ^{205}Bi, ^{206}Bi, ^{206}Po, ^{207}Bi are required [12]. Besides hydrogen and helium, other gases like xenon and krypton and volatile elements like mercury, cesium, iodine, bromine, and rubidium find their way to the cover gas system (CGS) of the reactor. The radio-toxicity increase several thousand times in LBE target system compared to a normal power reactor [13]. Toxic elements like Po and Hg have to be safely removed from the surface of LBE coolant before exhaling gases in atmosphere. This exerts need of production cross section data of the products along with their rate of absorption on surface before exhaling out. Selenium and tellurium produced along with polonium carry special attention to be handled very carefully for removal of polonium [14].

4.2.1.2 Polonium Problem

Production of polonium (Po) is the major problem in case of either Pb or Pb–Bi eutectic [11] to be used as the spallation target. Isotope $_{84}$Po^{210} is produced from beta decay of $_{83}$Bi^{210} in the one-step process of $_{83}$Bi^{209} (n, γ) $_{83}$Bi^{210} reaction. Also, $_{84}$Po^{208} is produced in beta decay of the product of $_{83}$Bi^{209} $(n, 2n)$ $_{83}$Bi^{208} nuclear reaction. Isotope $_{84}$Po^{209} is produced in two-step beta decay of the product of $_{82}$Pb^{208} (n, γ) $_{82}$Pb^{209}. In fact, $_{84}$Po^{210} is an alpha emitter and causes high radio-toxicity for a long time ($T_{1/2}$ being 138.38 days) particularly when LBE is used as the target + coolant system in an ADSS design. LBE as a coolant also induces radio-toxicity in large parts of ADSS. Gromov et al. [15] have calculated accumulation of radio-toxicity of three isotopes of polonium, i.e., $^{208-10}$Po after operation of LBE irradiated by 20 MW beam, and found that radio-toxicity is 1.11 PBq by ^{210}Po, 1.85 TBq by ^{209}Po and 0.13 TBq by ^{208}Po for a coolant mass of $\sim$10 tons.

Polonium rapidly combines with lead and forms a stable PbPo, and it becomes basis of removal by the following five processes [15, 16],

(i) Polonium hydride stripping using the following reaction after flowing hydrogen over the irradiated LBE,

$$PbPo + H_2 = H_2Po + Pb \quad (4.5)$$

H_2Po being a volatile helps in fast removal and Pb being constituent of LBE remains behind.

(ii) Direct distillation of PbPo at high temperature.

(iii) In alkaline extraction, irradiated LBE is reacted with NaOH to form Na_2Po which is separated,

$$PbPo + 4NaOH = Na_2Po + Na_2PbO_2 + 2\ H_2O \quad (4.6)$$

(iv) By the electrolysis, PbPo is deposited to separate it from LBE.

(v) Separation of polonium by the formation of solid polonide.

In fact, removal of Po from the irradiated 'target cum coolant' is a matter of mass scale extraction of Po, and all the aforesaid methods are under research and development. It is important to note that LBE has been used as coolant by Russian submarines however with no publication on its usage industrially.

LBE-cooled ADS systems have been proposed at Los Alamos National Laboratory for burning actinides and LLFP of spent fuel from light water reactor and another similar system [16].

4.2.1.3 Beam Window

Beam window is another part of the target system, and it works in between vacuum on one side and high-density and high-temperature material on the other side. According to Sugawara et al. [17], material of the beam window should possess the following characteristics of tolerance to severe conditions like (i) pressure of the LBE, (ii) heat generation by the proton beam, (iii) creep deformation at high temperature, (iv) corrosion in the LBE, and (v) radiation damage by neutrons from target and protons of the beam. Using the computational fluid dynamics, STAR-CD code, temperature difference between inside and outside the window has been calculated to vary in between 25 and 60 °C and it is a function of the beam shapes, Gaussian, parabolic, or flat. It is inferred that in case of T91 ferritic/martensitic steel (Mod. 9Cr-1Mo, with Nb, V, Ni, Mn, P, Ti, Al, Cu, S, Si, Co and N and C), parabolic beam profile is more suitable for the given ADSS design conditions [17]. Nuclear data of most of the aforesaid elements/material for incident proton or neutron up to 200 MeV energy is available in standard data libraries and beyond 200 MeV data can be available from the theoretical models or in some cases from the cosmic ray experiments.

Wang et al. [18] have studied graphene-beam-window design considering graphene (carbon dominant) which has high thermal conductivity, high strength, and high transparency to ions. According to a simulation study for a 10 MW proton beam, power spatial distribution of temperature of the window varies from 33 to 55 °C in case of Gaussian and 36.2–78.6 °C in case of square beam profile. The DPA distribution for the 251 $\mu A/cm^2$ peak current varies from 8 at the center to 3 on periphery in case of 2D Gaussian beam.

In this way, for the beam window, the following data is required,

(i) For the p + C, p + Fe, p + Cr, p + Mo, p + Nb, p + V and p + N and other similar reactions as given above, interaction data (elastic and inelastic cross sections, product yield, kinetic energy, and emission angle of the products, etc.) at beam energies varying from 1 to 1.5 GeV with the required accuracy is required. In the present situation of databases such as ENDF VII.0 [19], JENDL 4.0 [20], JEFF 3.1 [21], ROSFOND 2010 [22], TENDL-2009 [23], MENDL-2 [24], etc., data of all required proton energies is not available. One can make use of either the model calculations using TALYS [25], ALICE [26] or utilize the older database developed by Barashenkov et al. [27]. All the data of the database of Ref. [27] cannot be displayed here, and reader is advised to go through a part of the data displayed in Chap. 2 of Ref. [28].

(ii) Data related to creep rate and strength of the steel or any other perspective material of the beam window, as a function of (i) temperature and (ii) LBE flow speed in between 300 and 600 °C for the ADSS reactor is required. Creep resistance at high temperatures of steels has been studied extensively. For example, creep resistance can be increased by adding evenly distributed nanoparticles as traces in the volume or oxide layer [29, 30] on the surface. Creep rate as function of time at ~923 °K in case of boron disbursed in the 9% Cr ferritic steel has been studied [31]. According to Jianu et al. [32], T91 steel has been found to have better strain rate, lesser rupture rate, and rapid transition into the third creep stage at high stress (above 180 MPa) in the LBE environment than air. In Ref. [29], detailed study of the LBE technology is presented.

(iii) Several groups are engaged in calculating cooling of the beam window with the help of circulating LBE as spallation target and the coolant [33].

(iv) Radiation damage cross sections and other basic data such as threshold displacement energy, $E_{d,}$ of the window materials for Fe, Cr, Mo, V, C, N, O is required. Some of the already available data of alpha and ^{3}He gas production by Pb and Bi targets, available in standard data files is shown in Fig. 4.2.

4.2.1.4 Lead, Bismuth, and LBE as Spallation Target

High-energy proton beam after emerging from the beam window collides with bulk of a high Z and high N material which can be a solid, e.g., Pb, Bi, W, Ta, Hg, ^{nat}U, or Th, or a liquid in the form of an eutectic. Lead-bismuth eutectic (LBE) is widely accepted as a neutron source and coolant. LBE getting popularity over other solid targets particularly in case of application of hard neutron energy spectrum in systems like ADSS for transmutation of actinides. LBE has melting point 398 ± 1 °K with melting heat (latent heat capacity) 38.6 ± 0.3 kJ/kg and boiling point 1,927 ± 16 °K. This shows that LBE has high possibility of heat capacity and can be used as coolant for a system. High boiling temperature shows its high safety feature of elimination of high pressurization and other boiling concerns related to the core of the system. They have been the issues in case of coolants like Na, H_2O, D_2O, and He which have very low boiling points. This also prevents high reactivity effects from boiling. Additionally, both Pb and LBE are inert compared to Na as a coolant. In Table 4.2, physical parameters of Pb, Bi, and LBE recommended [29] for developing spallation target technology are presented.

In place of a solid, an eutectic is being considered because of its multiutility (i) possibility to utilize LBE as a coolant of the reactor as discussed earlier along with being used as a spallation target, (ii) having high thermal conductivity, (iii) lesser problems of changing the target shape, size, and physical behavior after long irradiation by HE beam, and (iv) high thermal capacity for carrying target heat and heat of the reactor. Its main snag is production of polonium (Po) on irradiation which is a neutron poison of thermal and fast reactors and a producer of high radio-toxicity. Other inherent impurities of Pb or LBE like Ag, Cd, Cr, Cu, and Fe, etc. affect thermal and hydraulic characteristics of the reactor. According to a data source [29], recommended data is largely satisfactory

Table 4.2 Recommended values and correlations of thermo-physical quantities, melting point, latent heat (L) of melting, boiling point, latent heat (L) of boiling, critical points, surface tension, density, sound velocity, thermal conductivity and dynamical viscosity of lead (Pb), bismuth (Bi) and molten lead-bismuth eutectic (LBE). Temperature, T, is expressed in Kelvin [29, 30]

Quantity (unit)	Pb		Bi		LBE	
	Data from a correlation	Temp. range (K)	Data from a correlation	Temp. range (K)	Data from a correlation	Temp. range (K)
Melting temp. (K)	600.6	–	544.6	–	398	–
L_Q melting (kJ kg^{-1})	23.07	–	53.3	–	38.6	–
Boiling temp. (K)	2,021	–	1,831	–	1,927	–
L_Q boiling (kJ kg^{-1})	858.6	–	856.2	–	856	–
T_c (K)	5,000	–	4,500	–	4,800	–
ρ_c (kg m^{-3})	3,250	–	2,800	–	2,200	–
P_C (MPa)	180	–	135	–	160	–
Surface tension (Nm^{-1})	$(525.9–0.113\ T) \times 10^{-3}$	601–1,300	$(420.8–0.081\ T) \times 10^{-3}$	545–1,400	$(448.5–0.08\ T) \times 10^{-3}$	420–1,400
Density (kg m^{-3})	11,441–1.2795 T	601–1,900	10,725–1.22 T	545–1,300	11,065–1.293 T	400–1,300
Sound velocity (ms^{-1})	1,953–0.246 T	601–2,000	$1{,}616 + 0.187\ T – 2.2 \times 10^{-4}\ T^2$	545–1,800	1,855–0.212 T	400–1,100
Thermal conductivity (Wm^{-1}K^{-1})	9.2 + 0.011 T	601–1,400	$7.34 + 9.5 \times 10^{-3}\ T$	545–1,000	$3{,}284 + 1.617 \times 10^{-2}\ T – 2.305 \times 10^{-6}\ T^2$	403–1,100
Dynamical viscosity (Pa s)	$4.55 \times 10^{-4} \exp (1{,}069/T)$	601–1,500	$4.456 \times 10^{-4} \exp (780/T)$	545–1,300	$4.94 \times 10^{-4} \exp (754.1/T)$	400–1,200

with respect to the experiments. Further, studies related to effects of impurities and dissolved additives on surface tension, sound velocity, viscosity, electrical resistivity, and thermal conductivity in case of Pb, Bi, and LBE need to be done to use as coolants.

Spallation process, neutron spectrum, and other characteristics of the neutron multiplication have been discussed in Chap. 3 extensively. As mentioned above, He^4 and He^3 gas production cross sections in spallation neutron colliding with Pb and Bi spallation targets have been shown in Fig. 4.2 [34–36]. In Refs. [34–36], data of other gas production is also given. Radiation damage cross sections of certain materials are given in next section.

4.2.1.5 Data Requirement for the Structure Material

Structure material like steel which is composed of several elements is used for containing LBE, circulation of coolant, and several other shielding structures. Similarly, for encapsulating and cladding of fuels as tube, covering of pallet or for a fuel-cell materials like aluminum, various carbides like ZrC, and SiC are used. They are expected to be irradiated by the beam protons or any such charged particle, produced neutrons, gamma, and heavy fission products in different positions in the reactor. They are strongly affected by the heat fluctuations and other hydraulic conditions. In case of irradiation by the high-energy neutrons of an ADSS, structure materials become source of gas production; i.e., gases like ${}_4He^2$ and ${}_2H^1$ affect the structure by way of swelling and embrittlement. Neutrons as well as other particles enhance the radiation damage of the structure materials. Radiation damage has been discussed in details in Chap. 7. Both fusion reactors and ADSS reactor are expected to experience a common conditions of high neutron fluence and temperature conditions. Additionally, the ADSS will experience effects of fluctuations of hydrodynamic flow along with jerks arising due to various conditions of solid fuels. The thermal and hydraulic stresses escalate corrosion and erosion of materials. All together, there is demand of data in this direction and in case of specific materials for the characterization, performance and efficiency calculations. It may be pointed out that even a middle energy reactor neutron ($\sim$1 MeV) produces helium gas on colliding with the Al-cladding or the Al-tubes of the fuel rods. Helium being inert gas, escapes out and produces bubbles to affect the dynamics of the coolants. Fusion reactors, having inner wall of vanadium alloy, are damaged by hard neutron spectrum as well as helium gas. Produced Helium if escaped toward the fusion core then it helps in enhancing the density in fusion chamber but on escaping in the alloy bulk of the inner wall it grows swelling like effects. In Fig. 4.3, displacement cross sections [37–40] of light materials, Al, V, Cr, and Ti are plotted while in Fig. 4.4 and He^4 and He^3—production cross sections of V and Zr [34–36] are plotted. These data are simulated data and need validation from experiments for which scientific community needs to evolve methods of measurement of displacement cross sections.

In Table 4.3, basic data of threshold damage energy, $E_{d,}$ of various target and structure materials [41] is given for a large number of Monte Carlo simulation works and experimental researches related to radiation damage. It may be recalled [42–44] that the thermal effects can influence the displacement energy because change of momentum reduces the recombination of correlated Frenkel pairs and lattice softening. Crystallographic direction plays a significant role [44, 45] in displacement energy and the calculation of the damage.

The requirement of data presented in this chapter needs a large amount of experiments and evaluation of the experimental data obtained from different sources for establishing compatibility and accuracy measures. The issue is discussed in Chaps. 2 and 4 by Chitra and Kumar [28] and in a large number of EXFORE workshops being held world over at the initiative of IAEA and the countries producing nuclear data. While discussing generic issues of a critical analysis of different data libraries, Ganesan has pointed out [46, 47] that in place of expected difference of $\sim$1% in ENDFB6GX and JENDL3GX of neutron absorption cross sections of ^{232}Th shows

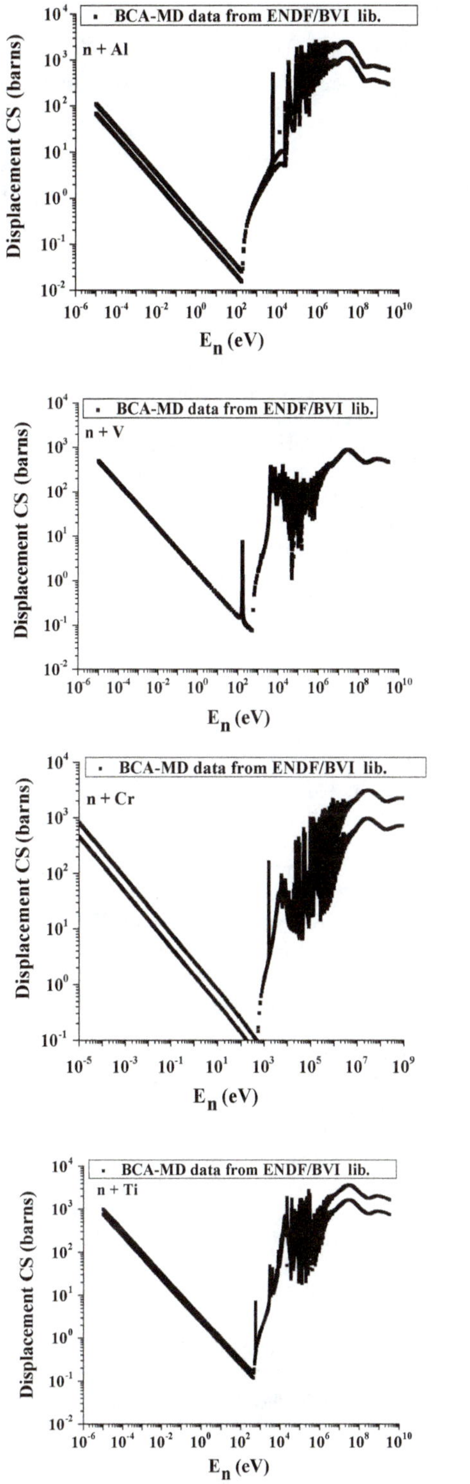

Fig. 4.3 Neutron displacement cross sections Al^{27}, V^{51} Cr^{24}, and Ti^{22} targets plotted as function of neutron energy [34–36]

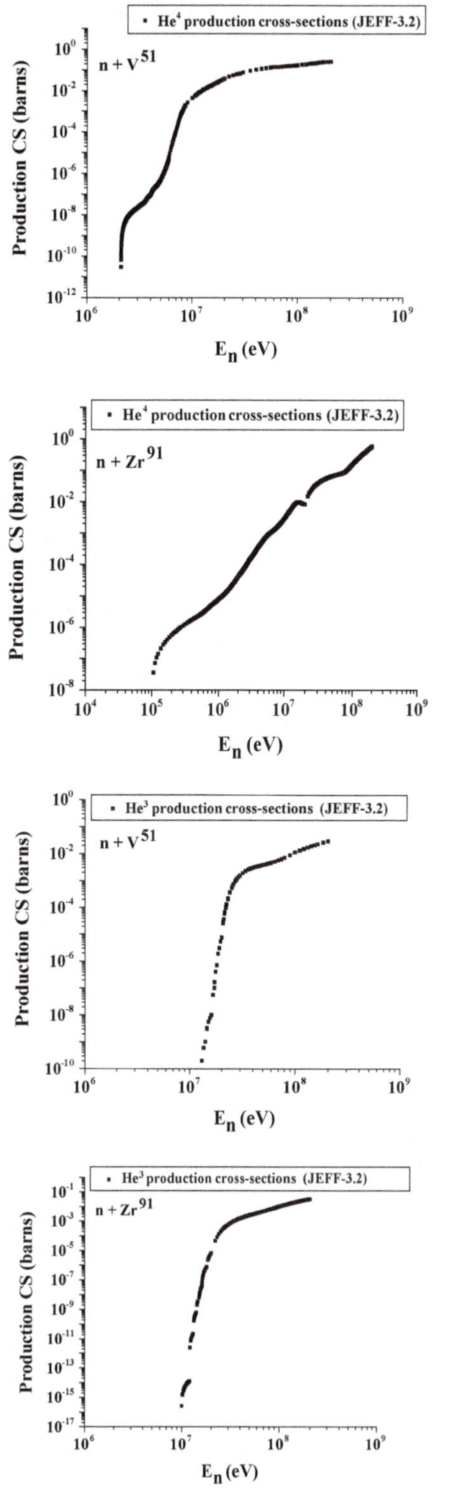

Fig. 4.4 He^4 and He^3 gas production cross sections in $n + V^{51}$ and $n + Zr^{91}$ collision of the specific structure materials plotted as function of neutron energy [37–40]

Table 4.3 Displacement threshold energy, E_d (eV) of target and structure materials [41]

Element	Al	Cu	Ni	Ag	Au	Pb	Pt	Pd	Th	Ge	Si	C 1	C 2	V
Lattice	FCC	FCC	FCC	FCC	FCC	FCC	FCC	FCC	FCC	Cubic	Cubic	Cubic	HCP	BCC
E_d (eV)	25, 27	29, 30	33, 40	39	40, 43	19–25	44	41	44	18–30	–	40	30	40
Element	Cr	Mn	α-Fe	Nb	Mo	Ta	W	Ti	Mg	Co	Zn	Zr	Cd	Re
Lattice	BCC	BCC	BCC	BCC	BCC	BCC	BCC	HCP	HCP	HCP	HCP	HCP	HCP	HCP
$\langle E_d \rangle$ (eV)	40	40	40–44	60, 78	60, 65	85,90	90, 100	30	20	36,40	29	40	30	60

C 1(diamond C), C 2(graphite C); lattice Ge, Si, C1 = diamond cubic

difference ranging from 0 to 600% at 293 °K. Similarly, comparison of fission cross sections of ^{233}U derived from JENDL-3.2 and ENDF/B-VI (Rev.5) files shows different magnitudes of difference at different temperatures. At 0 and 296 ° K, discrepancies between the two data are shown to vary between −89.60 to 1,495% and −88.44 to 597.9% respectively. Differences are observed in case of grouping of the data also. The differences have effect on calculations of k_∞ and critical masses.

References

1. Van Atta, C.M., Lee, J.D., Heckrotte, W: The electronuclear conversion of fertile to fissile material UCRL 52144, Oct 1976 and Van Atta, C.M.: A brief history of the MTA project. In: Proceedings of Information Meeting on Accelerator-Breeding, BNL Upton, New York, pp. 7–29, 18–19 Jan 1997. CONF-770107
2. Bartholomew, G.A., Tunnicliffe, P.R.: The AECL Study for an Intense Neutron Generator AECL-2600 (1966)
3. Knebel, J.U., Heusener, G.: Research on transmutation and accelerator-driven systems at the Forschungszentrum Karlsruhe. Internationale Zeitschrift für Kernenergie Atw Jg. (2000) Heft 6 (2000) 350
4. Ikeda, Y.: Nuclear Data Relevant to accelerator driven system. Nucl. Sci. Technol. **39**(Supplement 2), 13–18 (2002)
5. Plompen, A.: Why do We Still Need Nuclear Data. https://www.oecd-nea.org/science/meetings/pnd22/presentations/1-PLOMPEN.pdf
6. Koning, A.J., et al.: Nuclear data for accelerator driven systems. Nuclear models, experiments and data libraries. Nucl. Instrm. Meth. **A414**, 49 (1998)
7. Leray, S.: Nuclear reactions at high energy. Lectures Delivered at the Workshop on Nuclear Data for Science and Technology, Accelerator Driven Waste Incineration, Trieste (2001). http://users.ictp.it/~pub_off/lectures/lns012/Leray.pdf
8. MONC: Monte Carlo Nucleon Transport Code. http://bts.barc.gov.in/AuthStatic/auth_static.php/npd/monc/index.html
9. Serber, R.: Nuclear reactions at high energies. Phys. Rev. **72**(11), 1114 (1947) and Domengetroy, V.B.: Investigations Related to the Generation of Reaction Products in the Target of Accelerator Driven Systems for Nuclear Waste Incineration, FZKA 6908 (2003)
10. Parel, R.E., Lichtenstein, H.: User Guide to LCS: The LAHET Code System, Report LA-UR-89-3014 (1989)
11. Glasbrenner, H., et al.: Polonium formation in Pb–55.5Bi under proton irradiation, J. Nucl. Mat. **335**, 270 (2004)
12. Artisyuk, V., Saito, M., Sawada, T.: Current status of spallation product data: Nuclear engineering viewpoint. In: JAERI-Conference, p. 27 (2004–05). http://www.iaea.org/inis/collection/NCLCollectionStore/_Public/36/116/36116662.pdf
13. Accelerator-Driven Systems (ADS) and Fast Reactors (FR) in Advanced Nuclear Fuel Cycles. OECD NEA Report (2002). ISBN 92-64-18482-1
14. Amaya, L.O., Braet, J.: Purification of lead-bismut eutectic used in accelerator driven systems—9411. In: WM2009 Conference, Phoenix, AZ, 1–5 Mar 2009
15. Gromov, B.F., et al.: Liquid-metal lead-bismuth target for high-energy protons as an intense source of neutrons in accelerator-controlled systems. At. Energ. **80**(5), 378 (1996)
16. Buongiorno, J., Larson, C.L., Czerwinski, K.R.: Discussion on Polonium Extraction System for Pb-Bi cooled Reactors. Report FR0202114. http://www.iaea.org/inis/collection/NCLCollectionStore/_Public/33/031/33031171.pdf
17. Sugawara, T., et al.: Conceptual design study of beam window for accelerator-driven system, J. Nucl. Sci. Technol. **47**, 953 (2010)
18. Wang, H.: Design of high power graphene beam window. In: MOOCA 03, Proceedings of IPAC 2014, Dresden, Germany, p. 45. ISBN 978-3-95450-132-8
19. http://www.nndc.bnl.gov, http://www-nds.iaea.org/, http://www.nea.fr/dbdata/data/manual-endf/BNL-90365-2009.pdf
20. Shibata, K., et al.: JENDL -4.0: A new library for innovative nuclear energy systems. In: Proceedings

of the International Conference on Nuclear Data for Science and Technology, Jeju Island, Korea (ND-2010). http://wwwndc.jaea.go.jp/jendl/j40/j40.html. Accessed 26–30 Apr 2010
21. Jacqmin, R., et al.: Status of the JEFF-3 project. In: Proceedings of the International Conference on Nuclear Data for Science and Technology, vol. 1, p. 54 (2002). https://www.nea.fr/dbforms/data/eva/evatapes/jeff_31/
22. http://www.ippe.ru/podr/abbn/libr/rosfond.php
23. http://www.talys.eu/tendl-2009/. ftp:/nrg.eu/pub/www/talys/tendl-2009beta/Document/wonder09_rochman.pdf
24. Shubin, Y.N., et al.: Cross Section Data Library MENDL-2 to Study Activation as Transmutation of Materials Irradiated by Nucleons of Intermediate Energies. Report INDC (CCP)-385, International Atomic Energy Agency, May 1995
25. Koning, A.J., et al.: Proceedings of the International Conference on Nuclear Data for Science and Technology, Santa Fe, USA, 26 Sept–1 Oct 2004 and www.talys.eu/
26. Yamano, N.: Tables of isotope production cross-sections (ACSELAM Library). (Sumitomo Atomic Energy). http://wwwndc.jaea.go.p/ftpnd/sae/acl.html. Also see Fukahori, T.: ALICE-F calculation of nuclear data up to 1 GeV. In: Proceedings of the Specialists Meeting on High Energy Nuclear Data, Tokai, p. 114, 3–4 Oct 1991, JAERI-M 92-039 (1992)
27. Barashenkov, V.S., Toneev, V.D.: Interaction of High Energy particles and Nuclei with Nuclei. Atomizdat, Moscow (1972)
28. Bhatia, C., Kumar, V.: Role of (n, xn) Reactions in ADSS. Lambert Academic Publishing (2011)
29. Taneike, M., et al.: Creep-Strengthening of Steel at High Temperatures using Nano-Sized Carbonitride Dispersions. Nature **424**, 294 (2003)
30. Handbook on Lead-Bismuth Eutectic Alloy and Lead Properties, Materials Compatibility, Thermal-hydraulics and Technologies, NEA. https://www.oecd-nea.org/science/pubs/2015/7268-lead-bismuth-2015.pdf
31. Abe, F., et al.: Stabilization of martensitic microstructure in advanced 9Cr steel during creep at high temperature, Mater. Sci. Eng., A **378**, 299 (2004)
32. Jianu, A., et al.: Creep-to-rupture tests of T91 steel in flowing Pb–Bi eutectic melt at 550 °C, J. Nucl. Mat. **394**, 102 (2009)
33. Mantha, V., Mohanty, A.K., Satyamurthy, P.: Thermal hydraulic studies of spallation target for one-way coupled indian accelerator driven systems with low energy proton beam. Pramana J. Phys. **68**, 355 (2007)
34. https://www-nds.iaea.org/public/download-endf/DXS/Gas_Production_XS/DXS. (2015)/Files/
35. Konobeyev, A.Y., et al.: Evaluated displacement and gas production cross-sections for materials irradiated with intermediate energy nucleons. In: EPJ Web of Conferences, vol. 146, p. 02018 (2017). Accessed https://doi.org/10.1051/epjconf/201714602018
36. Konobeyev, A.Y., Fischer, U.: Complete gas production data library for nuclides from Mg to Bi at neutron incident energies up to 200 MeV. KIT Scientific Working Papers 36 (2015). ISSN: 2194-1629
37. https://www-nds.iaea.org/public/downloadendf/DXS/Displacement_XS/DXS3000. (2011) /Files/
38. Konobeyev, A.Y., Fischer, U.: Advanced evaluations of displacement and gas production cross sections for chromium, iron, and nickel up to 3 Gev incident particle energy. https://www-nds.iaea.org/public/downloadendf/DXS/Displacement_XS/DXS3000 (2011)/Description_of_methods/DXS_AccApp11.pdf
39. Konobeyev, A.Y., et al.: Improved displacement cross sections for structural materials irradiated with intermediate and high energy protons. https://www-nds.iaea.org/public/downloadendf/DXS/Displacement_XS/DXS3000.(2011)/Description_of_methods/DXS_AccApp07.pdf
40. Konobeyev, A.Y., et al.: Evaluated displacement and gas production cross-sections for materials irradiated with intermediate energy nucleons. In: EPJ Web of Conferences, vol. 146, p. 02018 (2017). Accessed https://doi.org/10.1051/epjconf/201714602018
41. OECD/NEA, Working Party on Multiscale Modelling of Fuels and Structural Materials for Nuclear Systems, Expert Group on Primary Radiation Damage. Primary radiation Damage in Materials. www.oecd-nea.org. NEA/NSC/DOC, 9 (2015)
42. Sosin, A., Bauer, W.: Atomic displacement mechanisms in metals and semiconductors. In: Dienes, G. J. (ed.) Studies in Radiation Effects in Solids, vol. 3, pp. 153–357. Gordon and Breach, New York (1969)
43. Drosd, R., Kosel, T., Washburn, J.: Temperature dependence of the threshold energy for Frenkel pair production in copper. J. Nucl. Mater. **69–70**, 804 (1978)
44. Roth, G., et al.: Energy dependence of the defect production at 78 °K and 400 °K in electron irradiated copper. Radiat. Effects **26**, 141 (1975)
45. Lucasson, P.: The production of Frenkel defects in metals. In: Robinson, M.T., Young Jr., F.N. (eds.) Fundamental Aspects of Radiation Damage in Metals. Oak Ridge National Laboratory, United States, pp. 42–65
46. Ganesan, S.: Nuclear data requirements for accelerator driven sub-critical systems—a roadmap in Indian context. Pramana J. Phys. **68**, 257 (2007)
47. Ganesan, S.: Nuclear data needed to develop new nuclear systems, role of n_TOF facilities to measure resonance cross-sections and nuclear data needs of thorium fuel cycle. In: Nuclear Physics and Astrophysics at CERN-NuPAC, Invited Oral Contribution, 10–12 Oct 2005

5 Transmutation of Spent Nuclear Fuel and Extension of a Fuel Cycle

To get rid of the danger of radio-toxicity of the radioactive nuclear waste, there can be several methods including reposition underground or in the space. For sending to the space, even the concept of Lagrangian points is under serious consideration. But all such reposition possibilities are not energy effective solutions because the waste contains a big amount of fissionable material also along with the fertile material that can be used for energy production. As a solution and from the point of reutilization of a bigger part of the unspent nuclear fuel (UNF), its reprocessing and transmutation are the most attractive solutions. However, there will always be a requirement of a solution where reprocessing is not necessary. As already discussed in Chap. 2 that hybrid fusion–fission systems may be consuming big lot of fertile fuels but the problem of long-lived nuclear waste (LLNW) will grow even by the hybrid systems. Thus, the world is looking forward for a better solution of the highly involved problem from the point of both reutilization and reduction of the LLNW by way of development of the accelerator-driven subcritical systems (ADSS).

Electronic Supplementary Material The online version of this chapter (https://doi.org/10.1007/978-981-10-7503-2_5) contains supplementary material, which is available to authorized users.

5.1 ADSS for Transmutation of LLNW

Natural transmutations by way of (i) radioactive decay and (ii) on collision of cosmic rays up to a deep inside the earth are well known. These observations have provided a way of transmutation by the artificial methods. In a power reactor, both transmutations by natural radioactive decay and by way of produced radiation like neutrons, gamma, and fission products take place. The two kinds of natural methods of transmutation are not sufficient because reactors produce much larger amount of UNF and HLW than they are reduced by natural methods. Transmutation by particles with sufficiently high energy fall under the category of artificially or the machine-induced transmutation. Historically, Alchemy has been used to convert lighter metals into a heavier metal hence a transmutation, but it has hardly reached to a respectable acceptance. After the advent of particle accelerators, artificial transmutation attained a scientific way of externally controlled method of activation.

In the spent nuclear fuel, there are actinides and fission products. Both have useful contents for applications as well as they pose danger of high radio-toxicity and risks of proliferation. Thus, the long-lived elements of the nuclear waste (termed as LLNW) are of serious concern from the point of their long survival time in reposition and otherwise they pose threats to the society during the reprocessing and transportation. A technology

V. K. Verma and K. Katovsky, *Spent Nuclear Fuel and Accelerator-Driven Subcritical Systems*, Green Energy and Technology, https://doi.org/10.1007/978-981-10-7503-2_5

for reduction of the LLNW can be appreciated when both energy is produced and it is degraded with respect to radio-toxicity and lifetime. This can be plausible in a system of very high neutron flux which can support both the energy production and the incineration.

5.1.1 Long-Lived Actinides (LLA) and the Fission Products (FP)

In a power reactor using uranium as fuel, isotopes of plutonium (Pu), neptunium (Np), and americium (Am) are among the main long lived actinides (LLA) produced in a reactor besides several isotopes of uranium. Radio-toxicity of several LLAs and FPs has been discussed in Chap. 1, and it has been observed that radioactive elements like Np, Pu, Am, and Cm as the TRU and Technician (^{99}Tc) and Iodine (^{129}I) are among the FPs having long half-life and they are highly radiotoxic also. In Table 5.1, half-life and decay mode of long-lived TRU and FPs are given. According to Wallenius [1], alpha-emitters MAs and TRUs are most radiotoxic and several of them are long-lived too. Several FPs are long-lived and their dose coefficient is equally high but they are relatively less radiotoxic as evident from the data given in Table 1.2 of Chap. 1. For the complete incineration of the long-lived nuclear waste (LLNW), a strategic planning is required to achieve the goal of degradation on one hand and gain of energy on the other hand.

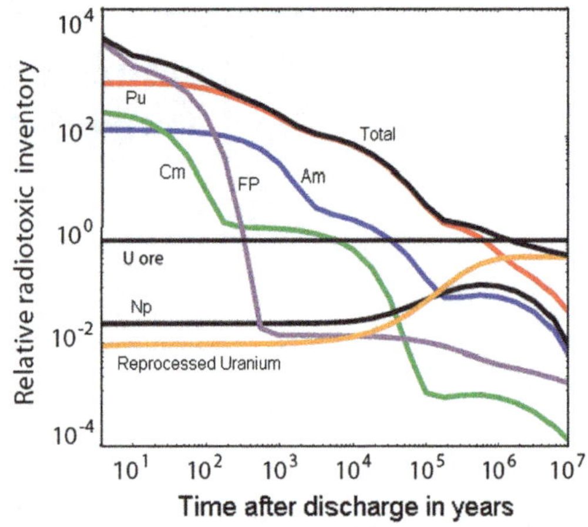

Fig. 5.1 Relative radio-toxicity of LLA and FP with respect to uranium ore in case of light water reactor (LWR) [2]

In Fig. 5.1, relative radio-toxicity (RRT) is plotted as a function of time after the discharge. From the figure, it can be inferred that radio-toxicity of FPs dominate other actinides up to 100 years and later on radio-toxicity of actinides dominates.

5.1.2 Reduction of Half-Life

Long-lived isotopes are a burden for several next generations, and this has been a matter of concern of the society and the nuclear science community in particular. Reposition is an alternative solution till an adequate system of

Table 5.1 Transuranium actinides (TRU) and fission products of a reactor having long life [1]

Element	Half-life (year)	Decay mode	Element	Half-life (year)	Decay mode
$^{236}Np_{93}$	1.54×10^5	ε(87.3), β(12.5), α(0.2)%	**^{79}Se**	3.27×10^5	β
$^{237}Np_{93}$	2.14×10^6	α	^{93}Zr	1.53×10^6	βγ
$^{238}Pu_{94}$	87.7	α	^{97}Tc	4.21×10^6	ε
$^{239}Pu_{94}$	2.41×10^4	α	^{98}Tc	4.20×10^6	β
$^{240}Pu_{94}$	6.564×10^3	α	^{99}Tc	2.11×10^5	β
$^{241}Am_{95}$	4.33×10^2	α	^{107}Pd	6.5×10^6	β
$^{242m}Am_{95}$	141	α(0.46%)	^{126}Sn	1.0×10^5	β
$^{243}Am_{95}$	7.37×10^3	α	^{129}I	1.57×10^7	β
$^{244}Cm_{96}$	18.1	α	^{135}Cs	2.3×10^6	β
$^{245}Cm_{96}$	8.5×10^3	α	^{79}Se	3.27×10^5	β

*Pu (Minor Actinide), Am (major or higher Actinide), and Cm (major or higher Actinide) are introduced in the fuel of the light water reactor (LWR)

Table 5.2 Typical nuclear reactions, half-life, $T_{1/2}$ (year) and spectrum average cross section (sp. av. cs.) of a product for the spallation neutron spectrum of 1 GeV p + Pb collision with preference of an incineration reaction shown in the last column

Reaction	Half-life of product ($T_{1/2}$)	Spectrum average cross section (mb)	Preferred reaction for incineration
Plutonium ^{239}Pu ($T_{1/2}$ = 2.41 × 10^4 year)			
^{239}Pu (n, γ) ^{240}Pu	6.56 × 10^3 y	40.78	
^{239}Pu (n, f)		1840	
^{239}Pu (n, 2n) ^{238}Pu	87.75 y	~130 mb	(n, xn)
^{239}Pu (n, 3n) ^{237}Pu	45.66 d		
^{239}Pu (n, 4n) ^{236}Pu	2.86 y		
^{239}Pu (n, 5n) ^{235}Pu	25.3 min		
^{239}Pu (n, 6n) ^{234}Pu	8.8 h		
^{239}Pu (n, 7n) ^{233}Pu	20.9 min		
^{239}Pu (n, 8n) ^{232}Pu	33.1 min		
^{239}Pu (n, 9n) ^{231}Pu	8.6 min		
^{239}Pu (n, 10n)^{230}Pu	1.7 min		
Neptunium ^{237}Np (T$_{1/2}$ = 2.14 × 10^6 year)			
^{237}Np (n, γ) ^{238}Np	2.117 d	189.35	(n, γ)
^{237}Np (n, f)	–	1372	
^{237}Np (n, 2n) ^{236}Np	1.54 × 10^5 y	~120 mb	
^{237}Np (n, 2n) ^{236m}Np	22.5 h		
^{237}Np (n, 3n) ^{235}Np	396.1 d		
^{237}Np (n, 4n) ^{234}Np	4.4 d		
^{237}Np (n, 5n) ^{233}Np	36.2 min		
^{237}Np (n, 6n) ^{232}Np	14.7 min		
^{237}Np (n, 7n) ^{231}Np	48.8 min		
^{237}Np (n, 8n) ^{230}Np	4.6 min		
^{237}Np (n, 9n) ^{229}Np	4.0 min		
^{237}Np (n, 10n) ^{228}Np	61.4 s		
Iodine ^{129}I (T$_{1/2}$ = 1.57 × 10^7 year)			
^{129}I (n, γ) ^{130}I	12.36 h	61.0	
^{129}I (n, 2n) ^{128}I	24.99 min	125.49	Both (n, γ) and (n, xn)
^{129}I (n, 3n) ^{127}I	stable		
^{129}I (n, 4n) ^{126}I	13.11 d		
^{129}I (n, 5n) ^{125}I	59.41 d		
^{129}I (n, 6n) ^{124}I	4.18 d		
Technetium ^{99}Tc (T$_{1/2}$ = 2.11 × 10^5 year)			
^{99}Tc (n, γ) ^{100}Tc	15.8 s		(n, γ)
^{99}Tc (n, 2n) ^{98}Tc	4.2 × 10^6 y		
^{99}Tc (n, 3n) ^{97m}Tc	90.6 d		
^{99}Tc (n, 4n) ^{96}Tc	4.28 d		
^{99}Tc (n, 4n) ^{96m}Tc	51.5 min		
^{99}Tc (n, 5n) ^{95}Tc	20.0 h		

incineration is developed. In case of incineration of actinides, possibility of energy gain by way of fission is a big point of attraction. Only a few of the minor products are long-lived which can be managed if less radiotoxic but difficulty arises in case of long-lived with high toxicity.

In an ADSS, maximum energy of the spectrum of spallation neutrons is very high and different compared to the neutron spectrum of a critical power reactor. Normally, it is called as a hard spectrum. In transport through the fuel, the spectrum is partly degraded in various nuclear elastic, inelastic, absorption reactions and fission. In Sect. 3.3.2, it has been shown that neutron spectra at different positions of an IAEA benchmark design are different when a neutron moves away from the spallation target toward the region 4 of the design. After emergence from region 3, spectrum is nearly a mixture of a thermal and a fast reactor. This provides opportunity to investigate possibility of degradation of long-lived nuclides by way of different kinds of nuclear reactions in the wide neutron energy range [3]. In Table 5.2, spectrum average cross sections of some of the nuclear products of a reactor are summarized with their half-life compared with the half-life of the parent nucleus (to be incinerated) in the environment of spallation neutrons. In turn, this suggests possible choices of degradation as mentioned above.

Besides the fission which is required for energy, ^{239}Pu when subjected to the spallation neutron spectrum, capture reaction cannot be a preferred reaction for incineration because its sp. av. cs. is much smaller than (n, xn) reactions and product of ^{239}Pu (n, γ) ^{240}Pu reaction also has half-life much longer than half-life of any product of (n, xn) reactions. In case of ^{237}Np, capture reaction is of choice due to high sp. av. cs. as well as much smaller half-life of its product than the products of (n, xn) reactions. On the other hand, ^{129}I can be degraded by both (n, γ) and (n, xn) reactions. In case of FP ^{99}Tc, capture reaction can be preferred due to lowest half-life of product ^{100}Tc.

In summary, from the point of reduction of half-life, one may have a choice of fuel like ^{238}U or ^{232}Th for the energy in an ADSS being designed having options of accelerating current, spallation target, and installing LLW in a specified area of fuel blanket for incineration looking at the availability of extra neutrons and their energy.

5.1.3 Reduction by the Thermal, Fast, and Spallation Neutron Spectra

^{239}Pu, ^{237}Np, and ^{129}I: For exploring possibilities of transmutation of ^{239}Pu, ^{237}Np, and ^{129}I by the thermal and fast components, and spallation neutron spectra itself of an ADSS reactor, calculated sp. av. cs. are compared in case of e.g., ^{239}Pu and ^{237}Np [3]. Data of sp. av. cs. for the fission and neutron capture reactions is given in Table 5.3 for the three neutron spectra. In case of ^{239}Pu, spectrum-averaged fission cross section σ_f is $\sim$257.0 b for thermal, 1.793 b for fast and 1.84 b for spallation neutron spectrum of ADSS. Apparently, a part of ^{239}Pu can be reduced to the FPs in a thermal reactor for the energy production. How the remaining amount of ^{239}Pu can be degraded by other nuclear reactions and it will depend on availability of extra neutrons. In case of a thermal reactor, the sp. av. cs. of ^{239}Pu (n, γ) ^{240}Pu reaction is $\sim$128.495 b and the product, ^{240}Pu itself has long half-life 6.56 $\times$ 10^3 year. Its neutron capture cross section $\sim$193.0 b is treated in the category of neutron poison of a thermal reactor. Another disadvantage of a thermal reactor is the high production rate of minor actinides by neutron activation, and most of the MAs are poorly fissionable. For example, the trans-plutonic nuclei such as ^{241}Am ($T_{1/2}$ = 432.0 years) and ^{243}Am ($T_{1/2}$ = 7370.0 years) are produced in the reactor on neutron capture and decay processes and they have relatively poor fission cross section $\sim$1.583 and 0.3436 b, respectively, in comparison of neutron capture cross sections, 220.47 and 43.96 b, respectively. Thus, they produce more amount of LLW like ^{242m}Am ($T_{1/2}$ = 142.0 years) and reduce the amount of neutron flux. Therefore, a thermal reactor can be said to be a neutron poorer and no extra neutrons will be available for further transmutation of other LLNW or FP such as ^{99}Tc and ^{129}I beyond the requirement of neutrons for energy production.

Table 5.3 Neutron economy factor (NEF) calculated for ^{239}Pu, ^{237}Np, and ^{129}I and the reaction rates, R in case of thermal, fast, and ADS reactors

Reactor and flux	Fuel/element	sp.av.cs., σ_f (b)	sp.av.cs., σ_c (b)	NEF, $(\sigma_f + \sigma_c)/\sigma_f$	Fission rate, R_f (s^{-1})	Capture rate, R_c (s^{-1})
Thermal, 8.9×10^{13} n/cm²/s	^{239}Pu	257.6	128.5	1.5	2.29×10^{-8}	1.14×10^{-8}
	^{237}Np	0.447	59.54	134.2	3.98×10^{-11}	0.53×10^{-8}
	^{129}I	–	7.53	–	–	0.57×10^{-9}
Fast, 1.69×10^{16} n/cm²/s	^{239}Pu	1.793	0.396	1.21	3.03×10^{-8}	0.67×10^{-8}
	^{237}Np	0.521	1.344	3.57	0.88×10^{-8}	2.27×10^{-8}
	^{129}I	–	0.14		–	0.24×10^{-8}
ADSS, 2.83×10^{16} n/cm²/s	^{239}Pu	1.84	0.041	1.02	5.21×10^{-8}	0.12×10^{-8}
	^{237}Np	1.372	0.19	1.14	3.88×10^{-8}	0.54×10^{-8}
	^{129}I	–	0.591[5]	–	–	1.67×10^{-8}

In a fast reactor, sp. av. cs. of capture reactions ^{239}Pu (n, γ) ^{240}Pu and ^{240}Pu (n, γ) ^{241}Pu are 0.396 and 0.445 b, respectively, and they are much smaller compared to a thermal reactor as discussed above. In case of spallation neutron spectrum, capture cross sections of the two reactions are even smaller, 40.87 and 76.08 mb, respectively.

5.1.3.1 Neutron Economy Factor (NEF)

Comprehensively, neutron economy factor (NEF) of a reactor can be defined by considering utilization of a neutron by fission alone or by the fission + capture reaction. It can be estimated by the ratio $(\sigma_f + \sigma_c)/\sigma_f$ as fission and capture reactions are different in cases of three neutron spectra. The ratio is (257 + 128.5)/257 = 1.5 for ^{239}Pu in a thermal reactor, (1.793 + 0.396)/1.793 = 1.21, for a fast reactor and (1.84 + 0.041)/1.84 = 1.02 for an ADS spallation neutron spectrum. Thus, ADSS requires least number of capture neutrons and can be considered as best for incineration of ^{239}Pu. Similarly, NEF in case of ^{237}Np for the three neutron spectra are given in Table 5.3 and compared with ^{239}Pu. It can be inferred that neutron economy is better in case of ADSS compared to both fast and thermal reactors for both ^{239}Pu and ^{237}Np individually. In case of ^{237}Np, NEF of a thermal reactor is about 120 times worse than ADSS and 40 times worse compared to a fast reactor due to capture cross sections being high.

5.1.3.2 Reaction Rate

A reaction rate $R = \sigma\Phi$ is directly proportional to the elementary interaction cross section for a given flux irradiating the fuel in a given volume. Reaction rate $R(A_r, Z_r)$ can also be defined in terms of production rate, $Q(A_r, Z_r)$ of a reaction product (A_r, Z_r) per target nucleus per incident projectile as $R(A_r, Z_r) = Q(A_r, Z_r)/\{N_t, N_{inc}\}$. Neutron flux of thermal and fast reactors are $\sim 8.917 \times 10^{13}$ n/cm²/s and $\sim 1.69 \times 10^{16}$ n/cm²/s, respectively, and in case of an ADSS, spallation neutron flux depends on the spallation target material, beam current and the beam particle. As discussed in Sect. 3.2.2, in case of 1 GeV p + Pb target system CASCADE code calculations show that on an average 24.3 spallation neutrons per beam proton are produced [4]. Thus, for 10 mA beam current or equivalently 6.25×10^{16} protons/s there are 1.51×10^{18} spallation neutrons/sec. Considering surface area of spallation target of dimensions $R \times L = 10$ 60 cm² (total surface area = 22,608 cm²), the spallation neutron flux $\sim 0.67 \times 10^{14}$ n/cm²/s. If we assume an ADSS reactor as shown in Fig. 3.6 corresponding to the situation that outer fuel cylinder of width 4 cm is filled with ^{239}Pu and $\langle \sigma_f \rangle$ of spallation neutrons is ~ 1840 mb, then the mean free path (m. f. p.) would be 10.86 cm. Thus, in the passage of 4 cm thickness of the fuel (Ref. Fig. 3.6), a spallation neutron makes ~ 0.3 fissions. If in a fission, $\nu \sim 1.9$ neutrons are produced then for every spallation neutron there are 0.57 neutrons produced in fission. Source neutron multiplication factor $\{k_s/(1 - k_s)\}$ = 0.57/(1–0.57) ~1.3 inside the so assumed Pu-reactor. Thus, the multiplied source neutron flux would be $0.67 \times 1.3 \times 10^{14} \sim 10^{14}$ n/cm^2/s. It is pertinent to point out that multiplication of spallation neutrons by the (n, xn) like channels and multiplication by the fission neutrons in an infinite volume of the fuel will further add to the neutron flux and total flux will be $\sim 10^{16}$ n/cm^2/s. Neutron multiplication in case of a fertile fuel by the non-fission reactions is several times stronger than a fissile fuel particularly in case of high-energy spallation neutrons, and it has been discussed in detail in Sect. 3.2.2. In Sect. 3.3.1, neutron flux is estimated for an IAEA hypothetical reactor using the CASCADE code. For the calculations of neutron flux of an ADSS, we have considered IAEA benchmark design as a model and calculated neutron flux at different interfaces of fuel regions of the benchmark. This data has been projected in Table 3.9. Corresponding to criticality, $k = 0.979$ neutron flux at interface '12' is 2.83×10^{16} n/cm^2/s in case of Th–U fuel and it has been used for all ADSS calculations given in Table 5.3. However, in case of pure ^{239}Pu fuel, neutron flux will be somewhat different. It is important to point out that fission rates, R_f of ^{239}Pu for the thermal, fast, and ADSS reactors are different only within one order of magnitude, i.e., 1.14×10^{-8}, 0.67×10^{-8} and $0.12 \times 10^{-8,}$ respectively. In case of ^{237}Np filled in a hypothetical thermal reactor, it is three orders of magnitude weaker compared to a fast reactor. Contrary to this, neutron capture rate R_c differs within one order of magnitude for both ^{239}Pu and ^{237}Np. Capture rate of ^{237}Np is within an order of magnitude in the thermal, fast reactor, and the ADSS reactors. However, it holds valid that the fission dominates in a fast reactor [5] over the thermal reactor. In case of ^{237}Np, the capture reaction ^{237}Np (n, γ) ^{238}Np followed by β^- decay and then the (n, γ) reaction converts ^{237}Np into a neutron poison, ^{240}Pu [6]. Thus, the chances of production of neutron poison in case of ^{237}Np fuel are more in a fast reactor than the two reactors. Capture rate of ^{239}Pu for the production of ^{240}Pu is lowest in case of ADSS; hence, it reduces risk of production of ^{240}Pu as a neutron poison in ADSS compared to fast and thermal reactors. On the other hand, fission rates of both ^{239}Pu and ^{237}Np are very high in ADSS than both fast and thermal reactors.

In a thermal and a fast reactor, the capture and fission reactions play a dominant role over other reactions. However, this understanding changes drastically in case of ADSS because of the presence of very high-energy neutrons. It may be pointed out emphatically that the actinides like ^{239}Pu when irradiated directly to the spallation neutron spectra then the NEF needs to be redefined by adding contributions of (n, xn) like reactions. Thus, the NEF can be redefined as, $(\sigma_f + \sigma_c + \sigma_{n,xn})/\sigma_f$. If we consider contributions of only $(n, 2n)$ and $(n, 3n)$ reactions in case of ^{239}Pu, then $\sigma_{n,2n} + \sigma_{n,3n} = 0.04 + 0.06 \sim 0.10$ b and NEF would be 1.07 in place of 1.02. In fact, contribution of all the (n, xn) type reactions further worsens calculation of NEF. Also, when a spallation neutron passes through the thick reactor fuel pile [7], it is slowed down and enhancement of capture cross section σ_c takes place which worsens the NEF value at moderated energies.

5.1.3.3 Transmutation Power of a Reactor, $P(A_r, Z_r)$

Transmutation by capture, fission, and other nuclear reactions has been discussed in detail in the preceding chapters. Choice of one such reaction depends on the nuclide that need to be transmuted. The transmutation power, $P(A_r, Z_r)$ is defined as the quantity of produced masses, $m(A_r, Z_r)$ per unit mass of the target, $m(A_t, Z_t)$ [8, 9]. In Ref. [3], $P(A_r, Z_r)$ is identified as the transmutation power and expressed in terms of the normalized activity, $a(A_r, Z_r)$ without accounting for the decay of a product, (A_r, Z_r) during irradiation.

$$P(A_r, Z_r) = \frac{A_r \cdot a(A_r, Z_r)}{\lambda(A_r, Z_r) m(A_t, Z_t) N_{avo}} \tag{5.1}$$

Alternatively,

$$P(A_r, Z_r) = R(A_r, Z_r) \cdot N_{INC} \frac{A_r}{A_t} \cdot t_{irr} \quad (5.2)$$

where

N_{INC}	intensity of beam (s^{-1})
A_r	mass of residual product nucleus
Z_r	atomic number of residual product nucleus
A_t	mass of target nucleus
t_{irr}	irradiation time (s)
$R(A_r, Z_r)$	reaction rate of product (A_r, Z_r)

Normalizing over 10^9 beam particles, transmutation power can be written as,

$$P_{norm}(A_r, Z_r) = \frac{10^9 P(A_r, Z_r)}{N_{INC}} \quad (5.3)$$

For comparison of any two reactors, one can select one or a few product residual nuclei for detection experimentally. Based on various experiments, using different proton and deuteron beams reaction rates of transmutation of ^{237}Np and ^{239}Pu are estimated and displayed in Table 5.4. From the Energy + Transmutation setup which is discussed in detail in Chap. 6, it is clear that the neutron spectrum irradiating the fuel elements is the spallation neutrons only in the beginning and it is moderated and multiplied after passing through the bulk of the fuel or the spallation target.

Integral number of beam protons, imparted on the spallation target is 0.88×10^{13}, 2.93×10^{13}, 1.10×10^{13} and 1.18×10^{13} for the 0.7, 1.0, 1.5 and 2.0 GeV beams, respectively [9]. Transmutation power of the four fission products ^{97}Zr, ^{132}Te, ^{133}I, and ^{135}I are estimated and presented in Fig. 5.2a–d for the ^{237}Np and ^{239}Pu fuel samples as a function of the proton beam energy, E_p.

From the results given in Fig. 5.2, it is clear that at all proton energies $P(A_r, Z_r)$ is always higher in case of ^{239}Pu than ^{237}Np and in case of ^{237}Np alone transmutation power is highest at 1 GeV proton energy. In case of individual FP, transmutation power of ^{239}Pu stays constant at proton energy > 1 GeV but higher than ^{237}Np. This shows that technologically transmutation power of plutonium is nearly independent of the proton beam energy > 1 GeV.

Considering results of ^{237}Np irradiated by neutron fluence from the two deuteron beams of the same setup given in Table 5.4, it can be inferred that they are similar to the results of proton beams at $E > 1$ GeV. This can be considered as an important result for designing an ADSS with p or d beam.

In another setup, GAMMA-3 [10] which is irradiated by the 2.33 GeV deuteron beam colliding with the Pb spallation target, transmutation power

Table 5.4 Comparison of measured sp. av. reaction rates, $\langle R(A_r, Z_r)\rangle$ in case of four proton beams of energies, E_p = 0.7, 1.0, 1.5, and 2.0 GeV and two 1.6 and 2.52 GeV deuteron beams imparted on ^{nat}U of $E + T$ setup in case of ^{237}Np. Reaction rates of FP from ^{239}Pu fuel are given in brackets [...] [9]. Measurement uncertainties are shown in brackets ( ) along with the data of reaction rates

Product	Reaction rates × 10^{-27}/s					
	0.7 GeV p	1 GeV p	1.5 GeV p	2 GeV p	1.6 GeV d	2.52 GeV d
^{238}Np	56.1(24)	151(5)	140(3)	133(3)	182(06)	162(06)
^{97}Zr	0.80(24) [5.52(51)]	2.10(09) [3.24(30)]	2.12(07) [8.82 (80)]	1.59(08) [11.35(10)]	1.98(12)	1.88(29)
^{132}Te	0.56(12) [4.18(17)]	1.79(18) [3.45(14)]	1.77(28) [4.59 (19)]	1.47(11) [13.81(40)]	1.61(12)	2.17(32)
^{133}I	0.83(40) [6.23(21)]	2.14(21) [6.08(21)]	2.01(24) [11.04(38)]	1.82(28) [16.94(15)]	2.00(22)	2.65(75)
^{135}I	1.52(22) [3.90(10)]	1.35(26) [4.86(12)]	2.36(28) [8.78 (22)]	1.96(18) [12.2 (4)]	2.21(22)	–

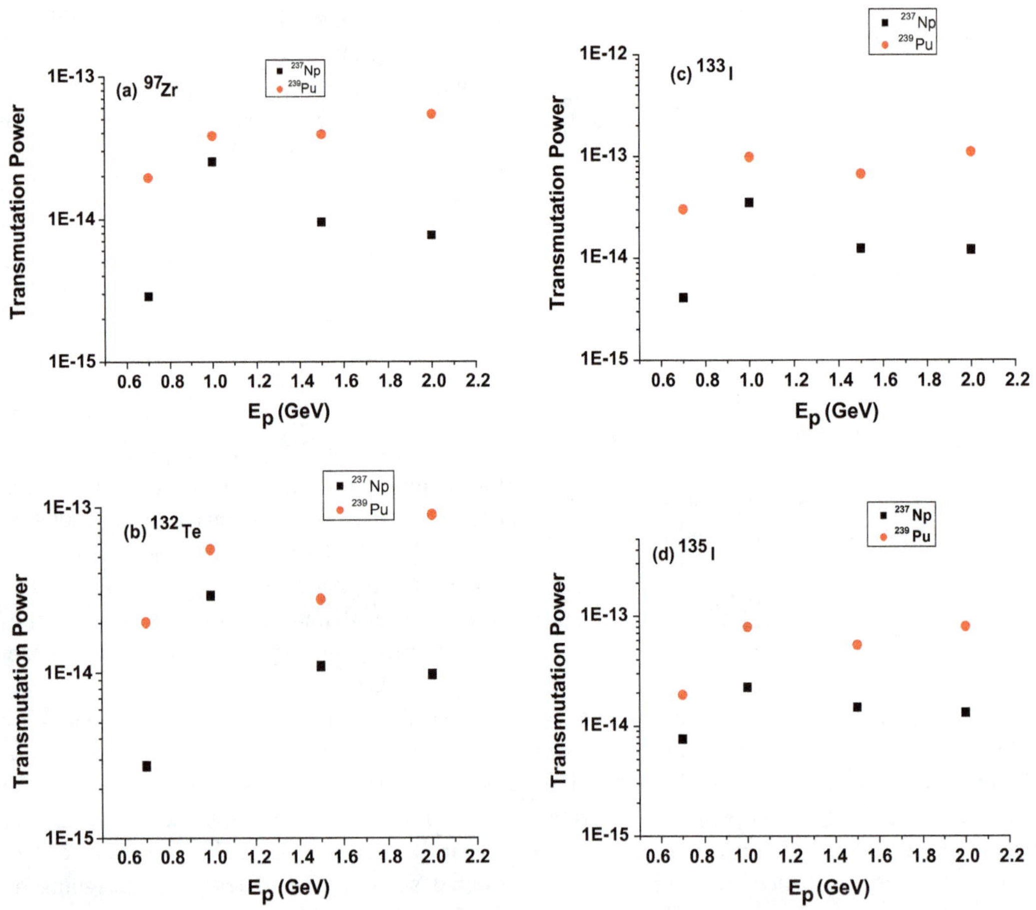

Fig. 5.2 Transmutation power, $P(A_r, Z_r)$ measured in the $E + T$ setup [9] for the four FPs, ^{97}Zr, ^{132}Te, ^{133}I, and ^{135}I of the ^{237}Np and ^{239}Pu fuel samples plotted as a function of proton beam energy, E_p

Table 5.5 Transmutation power of ^{232}Th and ^{nat}U measured at hole '*a*' of the Gamma-3 setup [10] when the setup is irradiated with 2.33 GeV deuteron beam with integral intensity $N_{INT} = 1.7(1) \times 10^{13}$

Produced Isotope	$\langle R(A_r, Z_r)\rangle$	$P(A_r, Z_r)$	Product	$\langle R(A_r, Z_r)\rangle$	$P(A_r, Z_r)$
^{232}Th:			^{nat}U		
^{233}Th	3.20(8)E −25	5.50E −12	^{239}Np	3.11(10)E −25	5.34E −12
^{85m}Kr	3.15(4)E −29	1.97E −16	^{85m}Kr	1.58(6)E −27	9.65E −15
^{99}Mo	1.99(19)E −28	1.45E −15	^{99}Mo	6.76(18)E −27	4.81E −14
^{131}I	1.69(59)E −28	1.63E −15	^{131}I	3.02(6)E −27	2.84E −14
^{133}Xe	3.00(20)E −28	2.94E −15	^{133}Xe	1.25(18)E −26	1.19E −13
^{135}Xe	1.06(14)E −28	1.05E −15	^{135}Xe	4.1(12)E −27	3.98E −14

in case of ^{232}Th and ^{nat}U samples are estimated using both the capture and fission products. Results of the analysis are presented in Table 5.5. From the data, it may be inferred that the GAMMA-3 setup can transmute ^{232}Th and ^{nat}U by the capture reactions almost equally. Transmutation power of fission process is about one order of magnitude higher in case of ^{nat}U than ^{232}Th. This is because of the fact that ^{nat}U has a component of fissile ^{235}U. Adam et al. [10] on comparing transmutation powers of ^{232}Th and ^{238}U by way of capture reaction in the *E* + *T*, GAMMA-3, and TARC setup [11] observed that transmutation power of GAMMA-3 setup is about an order of magnitude higher than the remaining two while the transmutation power of both *E* + *T* and TARC are at par equal. This is discussed in greater detail in Chap. 6.

5.2 Conversion of Fertile into Fissile and the Breeding Process

One of the earliest setups for conversion of fertile into fissile fuels like ^{239}Pu and ^{233}U by way of the spallation neutron spectrum produced by the deuteron beam was investigated in the year 1976 under the MTA program [12] of USA. We know that a power reactor works for the sustenance of a chain reaction, and it requires a specific amount of neutron flux. By neutron capture and beta becay, a definite amount of fissile fuel can be produced from a fertile fuel element. Natural fertile elements like ^{232}Th, ^{234}U, and ^{238}U can be converted to ^{233}U, ^{235}U, and ^{239}Pu, respectively, by neutron capture reaction (n, γ). On a neutron capture, artificially produced fertile elements of a reactor such as ^{240}U and ^{238}Pu can also produce ^{241}Pu and ^{239}Pu fissile elements. Similarly, capture of more than one neutrons in succession can convert a fertile element into fissile element. In this kind of conversion processes, transuranium fissile elements such as ^{245}Cm from ^{242}Pu, ^{239}Pu from ^{236}U, and ^{243}Cm from ^{241}Am are produced. The conversion of fertile into fissile fuel is a strong function of kind of neutron spectrum, e.g., thermal, fast, or spallation neutrons and the structure of the reactor from the point of availability of excess neutrons.

Absorbance of a neutron by a fissionable nucleus can give rise to (n, f), (n, γ), and (n, $xnyp$) types of inelastic reactions, thus

$$\sigma_a = \sigma_f(n,f) + \sigma_c(n,\gamma) + \sigma_{np}(n,xnyp) \quad (5.4)$$

Spectrum average cross sections calculated using TALYS code for both the fast and spallation spectra are given in Ref. [14]. For a breeder reactor, 'neutron production factor' η is defined as,

$$\eta = (\sigma_f/\sigma_a)\nu \quad (5.5)$$

ν is the number of fission neutrons. In case of thermal and fast reactors, roughly speaking, mainly fission and marginally (n, $2n$) reactions are responsible for the production of extra secondary neutrons and ν = 2.50, 2.49 and 2.97 for the ^{235}U, ^{233}U and ^{239}Pu fissile elements, respectively. Thus, in case of thermal and fast reactors only first two terms of the relation (5.4) are meaningful. In case of spallation neutron spectrum, there is a significant contribution of the third term, σ_{np} particularly in case of fertile fuel targets. Thus, factor ν needs an up-gradation accordingly. In Table 5.6, reproduction factor, η has been calculated for the fast and spallation neutron spectra for the three fissile elements, ^{235}U, ^{233}U, and ^{239}Pu assuming ν corresponding to only fission process of neutron multiplication. In fact, calculation of η is meaningful in case of a fast breeder reactor which produces more amount of fissile material on burning lesser amount of fissile material. In such breeders, conversion ratio (CR), breeding potential ($BR_{max} = \eta - 1$), and breeding gain (BG) are defined and they depend on the reproduction factor, η.

In case of spallation neutron spectrum, reproduction of neutrons depends on fission and other $n_{np}(n, xnyp)$ type reactions and η can be redefined as

$$\eta^* = \{(\sigma_f + \sigma_{np})/\sigma_a\}(\nu + \langle n_{np}\rangle) \quad (5.6)$$

In case of spallation, neutron spectrum contribution of $\langle n_{np}\rangle$ is meaningful in case of fertile elements and insignificant in case of fissile

Table 5.6 Reproduction factor, η for the fissile fuels, ^{235}U, ^{233}U, and ^{239}Pu fertile fuels, ^{232}Th and ^{238}U for the fast and spallation neutron spectra. All sp. av. cs. are in (mb) units

Fuel	Fast n-spectrum				Spallation n-spectrum			
	σ_f	$\sigma_c + \sigma_{n,2n}$	ν [13, 14]	η	σ_f	$\sigma_c + \sigma_{np}$ etc.	$\langle n_{np} \rangle$	η^*
^{235}U	2080	270 + 15.3	2.50	2.20	1184	106.4 + 141	–	2.31
^{233}U	2250	74.2 + 3.17	2.49	2.41	2209.7	41.8 + 66.01	0.11	2.45
^{239}Pu	1793	396 + 2.54	2.97	2.43	1840	40.78 + 129.4		~2.9
^{232}Th	7.29	391 + 7.85	2.60	0.05	35.1	146.1 + 154.01	13.82	9.26
^{238}U	11.4	317 + 25.1	2.07	0.07	72.4	116.8 + 139.8	~13	~9.71

elements because in case of thermal and fast reactors sp. av. cs. and average neutron multiplication σ_{np} and $\langle n_{np} \rangle$, respectively, are not significant. In Chap. 3, production term P has been computed by all kinds of reactions including fission reactions and projected in Table 3.6. In case of fertile, ^{232}Th total $P = 0.981$, $P_f = 0.15$ and on taking $\nu = 2.60$ and proportionate value $\langle n_{np} \rangle = 13.82$. For the given data in Table 5.6, $\sigma_f = 35.1$ mb, $\sigma_c = 146.1$ mb, and $\sigma_{np} = 154.01$ mb then the reproduction factor, η^* would be 9.26. This means that ^{232}Th will not always be triggered to fission or neutron capture but a large proportion produces (n, $xnyp$) reactions hence the reproduction factor is very high. When spallation neutrons pass through thic fuel, then the reproduced neutrons will have lesser energy and chances of (n, $xnyp$) type reactions will be reduced and neutron capture will rise drastically to 391 mb when the neutron spectrum will be close to the fast spectrum and to 5350 mb when it further retards to the thermal spectrum. Obviously, toward the end of the ^{232}Th-fuel, chances of neutron capture, σ_a are enhanced. Although this will reduce the value of η toward the end of the fuel but the rate of conversion of ^{232}Th into ^{233}U will be enhanced through the following route,

$$\mathrm{Th}^{232}(n,\gamma)\mathrm{Th}^{233} \xrightarrow[\beta^-]{} \mathrm{Pa}^{233} \xrightarrow[\beta^-]{} \mathrm{U}^{233} \quad (5.7)$$

In case of fissile ^{233}U in the spallation neutron spectrum, $P = 0.981$, $P_f = 0.94$, $\nu = 2.49$, and $\langle n_{np} \rangle = 0.11$ only. For the given data $\sigma_f =$ 2209.7 mb, $\sigma_c = 41.8$ mb, $\sigma_{np} = 66.01$ mb, then reproduction factor is calculated to be $\eta^* = 2.45$. In case of a ^{232}Th–^{233}U breeder, where ^{233}U will be highly produced due to high number of reproduction neutrons or the factor, η^* is high for the fertile ^{232}Th. The fissile ^{233}U will be consumed by way of the fission reaction and the reproduction factor in the fast spectrum reduces to $\eta = 2.41$ only. Resultantly, fissile ^{233}U will be produced at a faster rate than ^{234}U from the ^{233}U. One can easily calculate the breeding gain, BG = BR − 1 from the value of BR.

5.3 Extension of the Fuel Cycles

So far nuclear fuel cycle is meant for the complete cycle applications of a fuel for energy production, reprocessing, disposal, and refueling of a reactor after enrichment by fissile fuels. Normally, a part of the fertile and fissile fuels can be extracted from the unspent fuel as discussed in Chap. 1 in case of existing power reactors. In thermal and fast reactors there hold a ^{232}Th–^{233}U fuel cycle, for example, as shown in Fig. 5.3, where it can be seen that the fuel cycle is limited in between (n, $2n$) and a few (n, γ) reactions besides a few β- and α-decays. In case of spallation neutron flux, a large number of nuclear reactions take place as shown in Table 3.4 of Chap. 3. Several other nuclear reactions of a produced nuclide from ^{232}Th and ^{233}U irradiation in the ADSS reactor will take place and they are not shown in Table 3.4. Bhatia and Kumar [15] have attempted to bring out a ^{232}Th fuel cycle of the ADSS neutron spectrum as Fig. 5.4. From this following is worth discussion

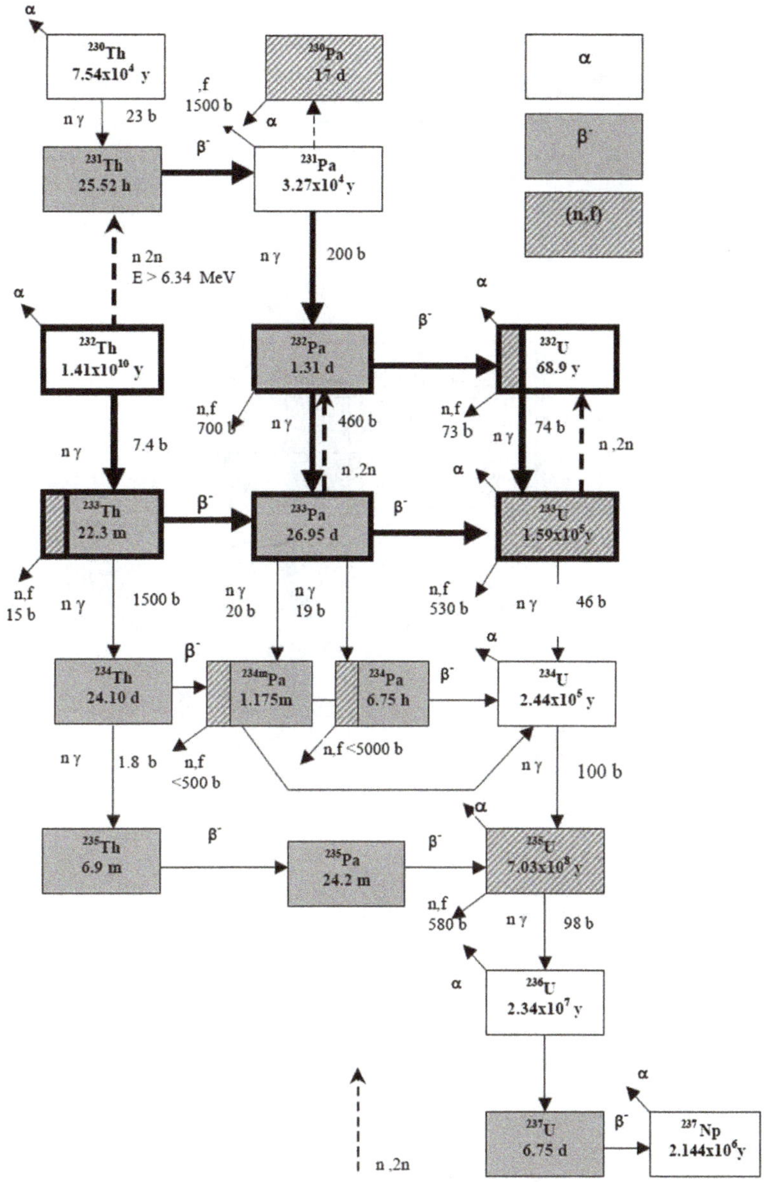

Fig. 5.3 The thorium chain, Th–U fuel cycle explored for thermal and fast reactors [16]

1. The fuel cycle in Fig. 5.4 is highly extended compared to the fuel cycle of a thermal and fast power reactor.
2. Neutron poisons of the thermal and fast reactors are no longer neutron poisons of an ADSS where spallation neutron environment is dominant compared to neutron spectra of critical reactors. This is also discussed in Chap. 3 of this book with data displayed in its Table 3.5.

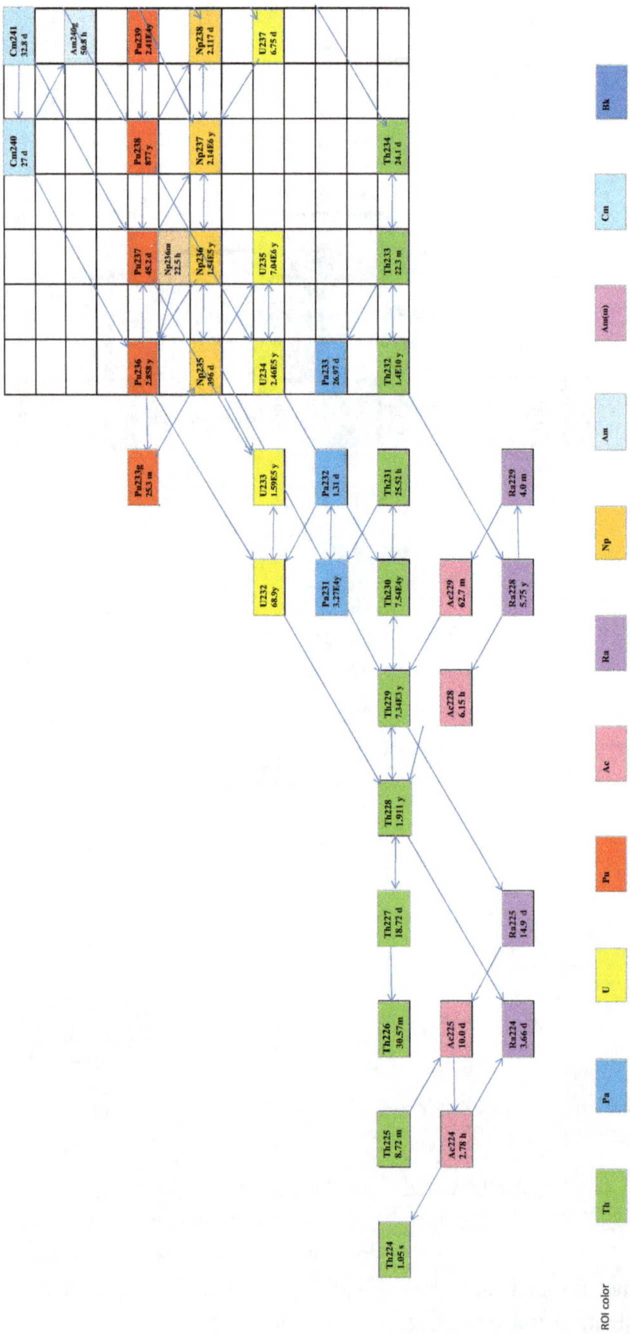

Fig. 5.4 (continued)

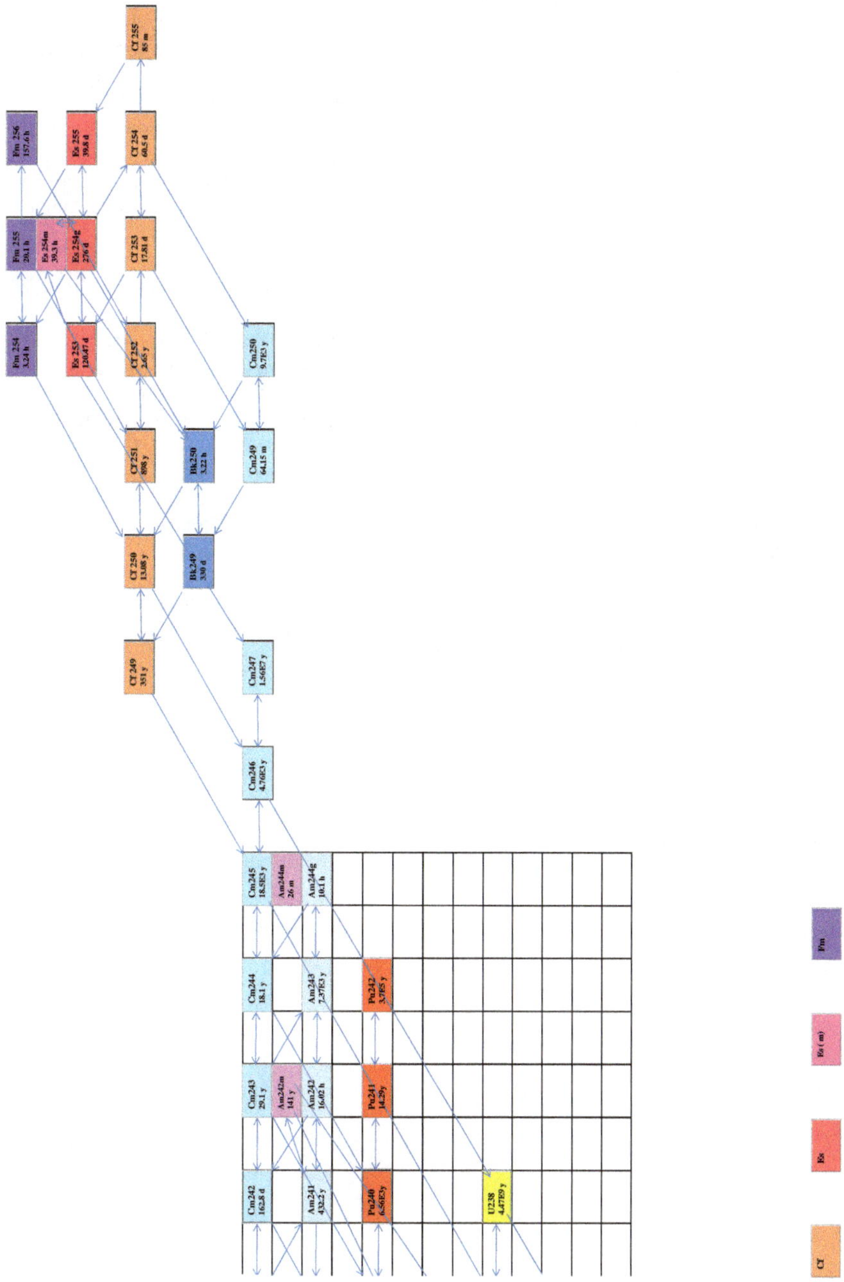

Fig. 5.4 Extended Th–U fuel cycle in the environment of high-energy spallation neutrons showing various possible reaction channels at high energy which are not feasible in a thermal or a fast spectrum [15]

References

1. Wallenius, J.: Transmutation of Nuclear Waste Reactor Physics. KTH, http://www.neutron.kth.se/courses/Transmutation.shtml and http://www.acsept.org/AIWOpdf/AIWO1-02-Wallenius.pdf
2. Development of reprocessing technologies. https://www.iaea.org/About/Policy/GC/GC52/GC52InfDocuments/English/gc52inf-3-att4_en.pdf
3. Kumar V. et al.: Study of Spent Fuel Management by Transmutation in an Accelerator Driven Sub critical System, EPE 2014, Brno Czech Republic, p 681, published as 978-1-4799-3807-0/14/$31.00 ©2014 IEEE
4. Kumar, V., Kumawat, H., Sharma, M.: Role of (n, xn) Reactions in ADS, IAEA-Benchmark and the Dubna CASCADE Code. Pramana: J. Phys. **68**, 315 (2007)
5. Junghans, A.R.: Transmutation of high-level radioactive waste: basics, methods, perspectives. In: Proceedings of Erinda-2 Colloquium, Dresden-Rossendorf, Germany (2013)
6. Kumar, V., et al.: A review of nuclear waste transmutation using accelerator beams up to several GeV: Talk delivered at 17th International Scientific Conference on EPE-2016, Praha (Czech Rep.) published as 978-1-5090-0908-4/16/IEEE (2016)
7. Sharma, M.: Ph.D. Thesis-Transmutation of Long-lived Isotopes of Conventional Reactors Using A.D.S. Concept. Ph.D. Thesis, University of Rajasthan, Jaipur, p. 177 (2010)
8. Adam, J. et al.: JINR Preprint P10-2000-28 (in Russian)
9. Adam, J., et al.: Transmutation of ^{129}I, ^{237}Np, ^{238}Pu, ^{239}Pu, and ^{241}Am using neutrons produced in target-blanket system 'Energy plus Transmutation' by relativistic protons. Pramana-J. Phys. **68**, 201 (2007)
10. Adam, J., et al.: A Study of reaction rates of (n, f), (n, γ) and (n, 2n) reactions in ^{nat}U and ^{232}Th by the neutron fluence produced in the graphite set-up (GAMMA-3) irradiated by 2.33 GeV deuteron beam. Eur. Phys. J. A **47**, 85 (2011)
11. Abanades, A., et al.: Results from the TARC Experiment: Spallation Neutron Phenomenology in Lead and Neutron-Driven Nuclear Transmutation by Adiabatic Resonance Crossing, Nucl. Instr. Methods **A478**, 577 (2002)
12. Van Atta, C.M., Lee, J.D., Heckrotte, W.: The electronuclear conversion of fertile to fissile material. UCRL 52144 (October 1976)
13. Turner, J.E.: Atoms, Radiation and Radiation Protection, 3rd edn. Wiley, New York (2007)
14. Bowman, C.D., et al.: Nuclear energy generation and waste transmutation using an accelerator-driven intense thermal neutron source. Nucl. Instrum. Methods Phys. Res. Sect. A **320**, 336–367 (1992)
15. Bhatia, C., Kumar, V.: Role of (n, xn) Reactions in Accelerator Driven Sub Critical Systems. LAP Lambert Academic Publishing (2011)
16. Thorium fuel cycle-Potential benefits & challenges. IAEA (May 2005), IAEA-TECDOC-1450, http://www-pub.iaea.org/MTCD/publications/PDF/TE_1450_web.pdf

6 Major Experimental Facilities for Development of Accelerator-Driven Subcritical System

Although the nuclear waste available in USA is in comparable amount that in Europe, yet, the EUROpean research program for the TRANSmutation of high-level nuclear waste in accelerator-driven systems (EUROTRANS) is initiated and funded heavily by the European Commission within its 6th Framework Program and it involves more than 40 partners which include research agencies, universities and nuclear industries. Initially, the program was planned for (i) advance design of a transmuter including its components including conceptual industrial design (ii) developing coupling of accelerator with the reactor (iii) studies related to advance fuels for transmuters (iv) studies and investigation of structure materials and heavy liquid metal technology and (v) collection of nuclear data with required precision.

In the following chapter, short description of all major experiments conducted world over in this direction, predated or postdated to EURO-TRANS, is presented. To start with most of the experimental facilities and the setups have utilized existing experimental and calculated data from the models and codes and reached to an initial stage of realization. It is inferred that there is need of new data for several materials with better precisions and this is discussed in Chap. 4. There is a need of development of new materials for high energy radiation and for several strategically suitable situations as well.

In the following, several facilities emerged world over for pursuing experiments to fulfill requirement of the EUROTRANS are described.

6.1 FEAT and TARC Experiment

FEAT is short form of '**F**irst experiment for testing energy gain in **E**nergy **A**mplifier (EA) and **T**ransmutation' as proposed by Nobel Laureate, Carlo Rubia in the year 1994 [1]. Later in this chapter, it will be seen that several experiments are conducted mainly for the reduction of average life of LLNW and hardly anyone for the study of energy production from the fertile component of the nuclear waste. Among several fission products which have long life and high radio-toxicity, ^{99}Tc and ^{129}I can be transmuted to reduce to either very low half-life or to become stable by way of neutron capture or other reactions. Looking at the properties of spallation neutrons on passing through lead (Pb), Rubia [2] proposed a new concept that neutrons in **A**diabatic **R**esonance **C**rossing can be utilized for absorption in a long-lived fission product (LLFP) like ^{99}Tc ($t_{1/2} = 2.11 \times 10^5$ yr) for their **T**ransmutation. It is abbreviated as TARC. Spallation neutrons after multiplication in a large lead block will have almost isotropic elastic scatterings and a 1 MeV neutron will suffer nearly 1,800 elastic collisions in its traversal length of about 60 m inside lead, and it attains kinetic energy

V. K. Verma and K. Katovsky, *Spent Nuclear Fuel and Accelerator-Driven Subcritical Systems*, Green Energy and Technology, https://doi.org/10.1007/978-981-10-7503-2_6

< 1 keV which lies in the resonance energy region. At 5.6 eV energy, it can be allowed to be absorbed in ^{99}Tc with a cross section ~4000 b compared to the cross section being 20 b in case of thermal and epithermal energy ≤1 eV. In fact, integral absorption cross section in the resonance region is ~310 b, again a very high cross section compared to the thermal, epithermal, and the high-energy regions. According to the concept of 'transmutation by adiabatic resonance crossing' (TARC), a neutron is contained in the large assembly for about 3 ms and finally it has high chance of absorption in ^{99}Tc to transmute it to become ^{100}Tc ($t_{1/2}$ = 15.8 s) which finally decays to ^{100}Ru (stable).

The TARC setup is an assembly of rectangular lead (99.99% pure) blocks assembled finally in cylindrical manner with size $d \times l$ = 3.3 × 3 m^2, and its axis is aligned to the beam axis. Its weight is ~334 tones. Nearly 70% spallation neutrons produced by the proton beam on striking lead material inside a beam hole at the center of the assembly remain inside the assembly and in an area of radius 1 m, a region is produced which is least affected by the back ground neutrons reflected from the atmosphere. This is done by keeping the concrete shielding far away from the setup. The setup is protected from the reflected neutrons from the ground also by introducing 44-cm-thick steel in between the setup and the ground and introducing 3-cm layer of B_4C-rich cement in between, for minimizing thermal neutrons. In Fig. 6.1, TARC setup is shown with mention of its dimensions [3]. For the FEAT experiment, 3.5 GeV/c proton beam extracted from the standard PS of CERN was used. It was a bunched beam having intensity 3 × 10^7–2 × 10^{10} particles per shot.

FLUKA code [4] is used to simulate the spallation neutron spectrum and transport of neutrons up to 19.6 MeV, and histories are recorded in a file. At higher than 19.6 MeV energies, transport is performed using the EA Monte Carlo simulation code specially written by the TARC group at CERN [5]. In fact, TARC experiment was conducted for the validation of the EA code like that CASCADE code was developed and validated by conducting experiments with GAMMA and $E + T$ setups at JINR, Dubna. They will be described later in this chapter.

Neutron database compilation [6] which carries several data files like ENDF VI-4, JENDL-3.2, JEF-2.2, EAF 4.2, CENDL-2.1, and

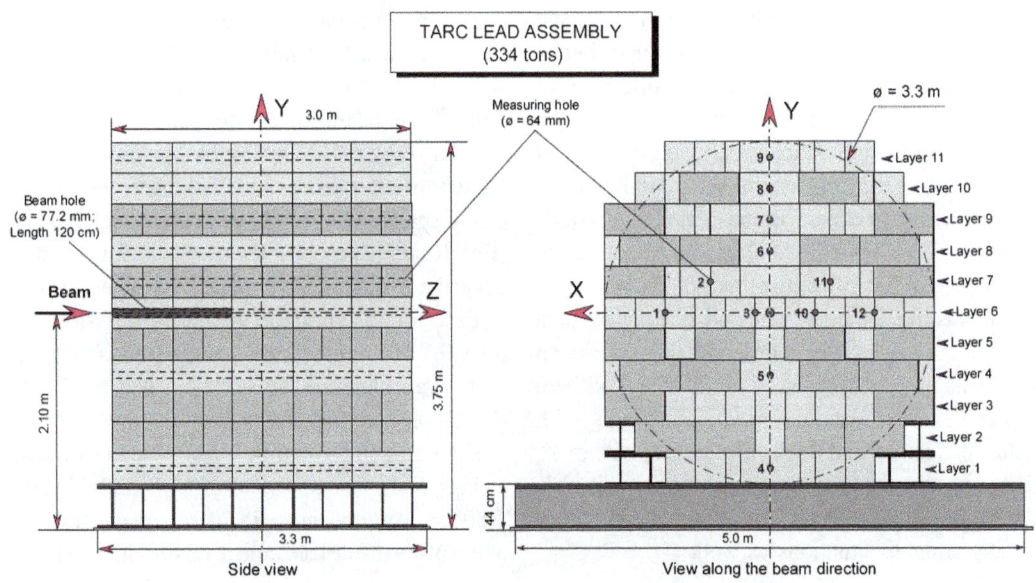

Fig. 6.1 TARC setup with mention of dimensions [3]

BROND-2 for different energy ranges and for specific materials has been utilized to prepare flat cross sections. It is a state-of-the-art procedure of utilization of data files after checking inconsistencies and after processing the raw data by the PREPRO code.

The TARC experiment takes care of the inaccuracies of the nuclear data and the material impurities whatsoever small, in the lead bulk. The setup has 12 holes for instrumentation and one big hole ($d \times l = 7.72 \times 120$ cm^2) for the beam-related procedures. In Fig. 6.2, neutron fluence measured using the ^{3}He neutron detector in one of the hole positions is plotted with the simulated data of the dedicated EA code for the 2.5 GeV/c proton beam colliding the setup.

Transmutation rate of ^{99}Tc is deduced from the gamma from de-excitation of ^{100}Tc to ^{100}Ru. In Fig. 6.3, decay schemes of the three nuclei are given. By way of detection of ^{100}Ru by the gamma spectroscopy, rate of decay of ^{100}Tc is deduced and it can be verified from its built up rate from neutron activation of ^{99}Tc by the given neutron spectrum in the activation position.

Similarly, experiments of capture rates of long-lived fission products ^{129}I and ^{127}I are studied in the TARC setup. Also, production rates of ^{233}U from ^{232}Th and ^{239}Pu from ^{238}U are estimated in the setup. Some of its results of actinides will be discussed along with the results of GAMMA-3 and $E + T$ experiments later in the chapter.

6.2 n_ToF Experiments for Measurement of Cross Section

Abbreviation *n*_ToF is commonly used for 'neutron time of flight.' The time of flight concept is applicable for determination of kinetic energy, E of an object of mass, m. Thus, for a neutron of mass, m moving with non-relativistic speed, v, its kinetic energy, E_n (eV) can be measured in terms of time t (μs) taken in flight length, L (m) using the following relation after rationalization of units,

$$E_n = \frac{1}{2}mv^2 = \left(72.3 \times \frac{L}{t}\right)^2 \quad (6.1)$$

This leads to energy resolution, $(\Delta E/E)$,

$$(\Delta E/E) \sim 2 \times \Delta t/t \quad (6.2)$$

In case there can be uncertainty in measurement of length, L also then Eq. (6.2) is written as,

$$\Delta E/E = 2 \times \sqrt{\left[(\Delta t/t)^2 + (\Delta L/L)^2\right]} \quad (6.3)$$

In this case, position of start is defined by $L = 0$ and $t = 0$, and they are assumed to be measured precisely. If there is any difference in time of measurement and time of start of the projectile then corresponding correction will be required. As we know spallation neutron sources

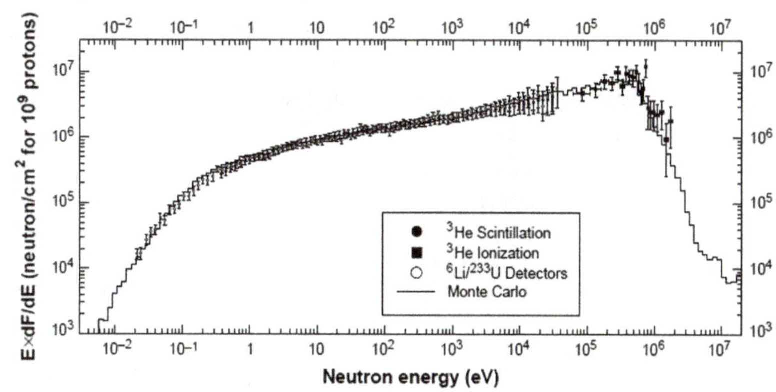

Fig. 6.2 Neutron fluence measured at $z = +7.5$, hole number '10' using the ^{3}He scintillator, ionization chamber, and ^{6}Li/^{233}U detectors (different for different *n*-energy ranges) described in Ref. [3] and the calculated neutron fluence by using the MC code

Fig. 6.3 Decay schemes of ^{99}Tc, ^{100}Tc and ^{100}Ru

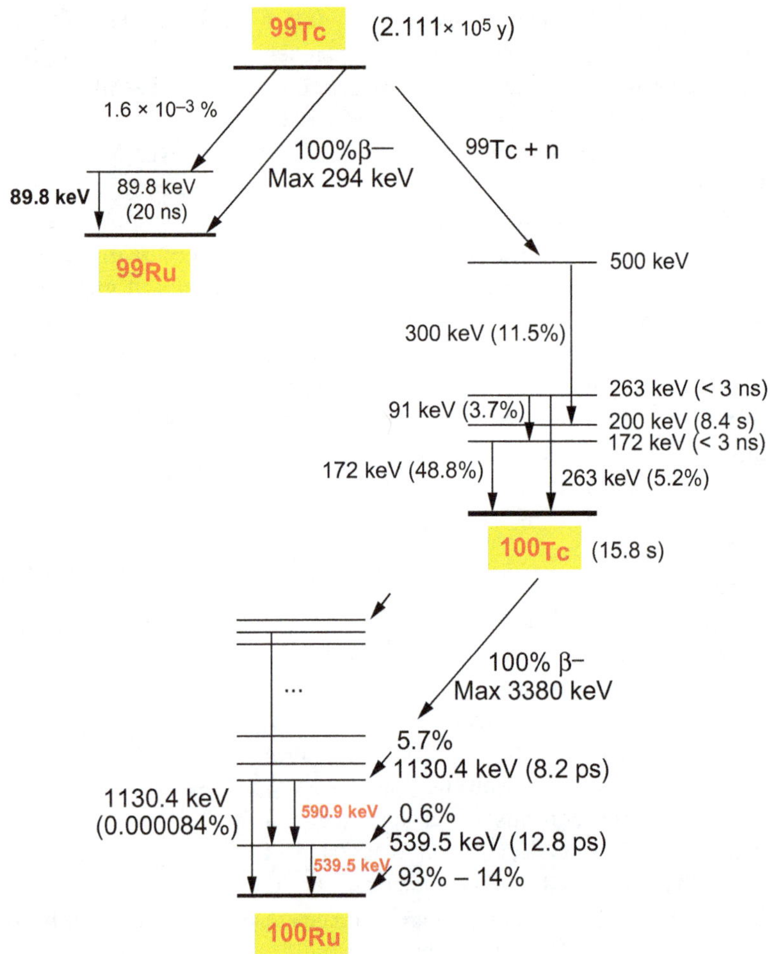

are driven by the accelerated particle beam, may be electron, proton, or deuteron, etc., therefore, pulse width will contribute toward Δt. For example, if there is a pulsed source of 100 ns width and a neutron time of flight is 11 m, then attainable energy resolution is $\sim 11.2\%$. Obviously, smaller pulse width and longer flight length, L, are most helpful for attaining better energy resolution which is required in resolving the highly narrow resonance peaks of cross sections in resonance region.

When a neutron is scattered by a nucleus on its way or it may have a nuclear reaction, then the scattered intensity I and incident intensity I_0 are related with the macroscopic cross section of scattering, $\Sigma = \sigma \times n$, by the following relation,

$$\frac{I}{I_0} = e^{-\sigma n x} \tag{6.4}$$

where n = number density and x = thickness in direction of the incident particle.

Thus, the total cross section σ can be written as

$$\sigma = 1/(nx) \times \ln\left(\frac{I_0}{I}\right) \quad (6.5)$$

Here, I_0/I is the inverse of the transmission coefficient, $1/T$ of the scattering material, and its measurement provides total cross section σ of the material. For having significant results of measurement of resonance cross section, it is essential that number of scatterings in the pure sample must be very large. This is possible with highly intense neutron beam and/or large thickness of the scattering material. Before starting an experiment, thickness of a sample needs to be optimized so that several secondary scatterings of incident particle do not take place in the given sample.

For a material composed of several isotopes, denoted by j, denominator of Eq. (6.5) can be substituted by $\Sigma_j N_j$ in place of $\Sigma_j n_j x$, and then the total cross section in the ith energy group can be written as,

$$\sigma(E_i) = \frac{1}{\Sigma_j N_j} \ln\left(\frac{1}{T_i}\right) \quad (6.6)$$

Wang et al. [7] have described the method of measurement of transmission coefficient using the Pohang Neutron facility (PNF) with details.

A major facility of cross-section measurement, the n_ToF facility at CERN, started in the year 2001 having 20 GeV/c pulsed proton beam (each pulse carries 3–7 $\times$ 10^{12} protons) with pulse width of 7 ns is allowed to fall on massive Pb target to produce spallation neutrons. On an average, 300 n/p are produced for the given beam power [8]. Other charged particles produced in a collision are removed from the beam line by the 1.5 T sweeping magnet. According to Guerrero et al. [9], neutrons produced from the spallation target are moderated by borated water tank to give a desired neutron spectrum ranging from 0.025 eV to 1 GeV energy. Energy resolution is $\Delta E/E = 10^{-4}$. The neutrons are collimated two times in an vacuum evacuated length of $\sim$182 m and to converge into a beam spot of 2–8 cm diameter; then they are allowed to enter in the experimental hall of about 7.9 m length. In several experiments, liquid C_6D_6 scintillation detectors are employed in two modes of prompt gamma detection and the neutron detections. Its setup is shown in Fig. 6.4. One can refer to Tobias et al. [9] for the details of the detector. In several experiments at CERN n_ToF, capture reaction yield is determined using Eq. (6.7). Here, $Y(E_n)$ is a function of measured total (σ_T) and capture (σ_γ) cross sections that are determined by peak fitting to the transmitted intensity data of experiment.

Capture reaction yield is given by the following relation,

$$Y(E_n) = \mu(E_n)\left(1 - e^{-n\sigma_T(E_n)}\frac{\sigma_\gamma}{\sigma_T}\right) \quad (6.7)$$

Here, μ corresponds to multiple scattering correction, σ_γ and σ_T are the capture and total cross sections, and they are functions of E_n-energy and determined by R-matrix formalism.

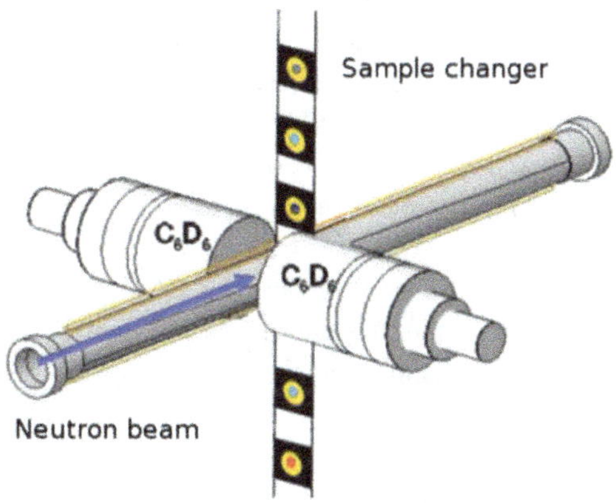

Fig. 6.4 Two C_6D_6 scintillation detectors ($d \times l$ = 12.7 $\times$ 7.62 cm^2) placed across the neutron beam and above a sample changer. The samples are put on kapton (canning) foils spread on carbon sheet to avoid n-back scattering. Each detector carries about 1 L of liquid scintillator (see Tobias et al. [9])

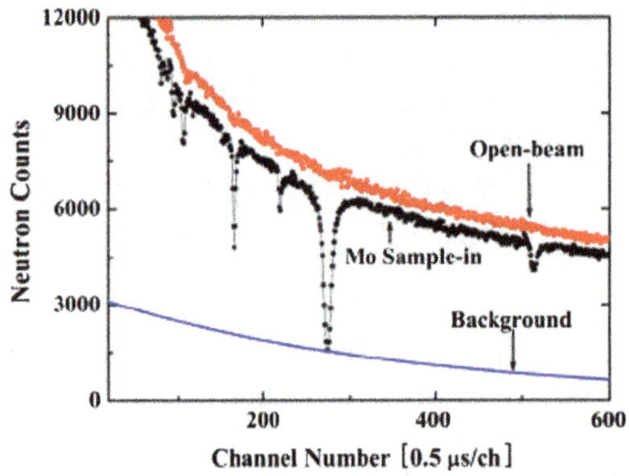

Fig. 6.5 Neutron counts versus the channel number in case sample in or open. Background spectrum is also projected [7]

In South Korea Pohang Neutron Facility, PNF [7] normally uses a large diameter ^{6}Li–ZnS(Ag) scintillator (BC702) for the neutron detection by way of ^{6}Li (n, α) ^{3}H reaction where both α and ^{3}H charge particles contribute toward generation of signal on falling on the ZnS scintillator. The signals are converted to electric current by a PM tube. In the facility, two data-acquisition systems are used, e.g., NIM for n-γ separation and CAMAC for the sample changer device and for the digitization of the 'start' and 'stop' signals. In Fig. 6.5, transmission spectrum of Mo sample taken at the PNF is shown along with the background counts. Reversed peaks correspond to the existence of resonances.

For the determination of resonance parameters, transmission spectrum is converted to cross-section versus neutron energy plot. Parameters are deduced from the SAMMY code [10] wherein multilevel R-matrix was used with Reich–Moore approximation [11] using the following relation,

$$\sigma_{\mathrm{T}} = \frac{2\pi}{k^2} g \left\{ 1 - \cos 2\phi \left(1 - \frac{\Gamma_n \Gamma}{2d} \right) - \sin 2\phi \frac{\Gamma_n (E_\lambda - E)}{d} \right\}. \quad (6.8)$$

E is the neutron energy and E_λ is the resonance energy, g is the statistical factor, $g = \frac{2J+1}{2(2I+1)}$ and width, $d = \left[(E_\lambda - E)^2 + \left(\frac{\Gamma}{2}\right)^2 \right]$, here, J is the spin of the resonance state and I is spin of the target nucleus, ϕ is potential scattering phase shift, and $\Gamma = \Gamma_\gamma + \Gamma_n$ is the sum of gamma and neutron widths expressed in eV. In practice, values of E_λ, g, J and Γ's are accessed from the theoretical data given by Mughabghab [12] for a large number of nuclei and their isotopes. Full width at half maxima, Γ can be determined from the resonance peak and the peak height, σ_0, related by the relation, $\sigma_0 = \frac{4\pi}{k^2} \frac{g\Gamma_n}{\Gamma}$ which allow determination of value of $g\Gamma_n$. In Table 6.1, data of measurements at CERN n_ToF and PNF are reported with references of several works shown in the table.

Derrien et al. [20] have presented results of transmission experiments conducted from the year 1972–1990 using ToF facility at Oak Ridge Electron Linear Accelerator (ORELA) including some more recent measurements. Most of the experiments have shown uncertainty more than >5% up to even 10%. In these measurements, Doppler and resolution-broadened capture cross sections were also calculated from the resonance parameters including self-shielding and multiple scattering effects. The n_ToF facility at CERN is world's highly placed facility of modern time and has longest path length ~185 m, and it has high capability of cross-section measurements with better accuracy. In particular, the facility has possibility of using very thin samples and precision is ~6% in case of averaged cross sections and <3.5% in case of resonance cross sections. Using the facility at CERN, González et al. have reported [22] preliminary results of a large number of resonance cross sections in the form of CS versus E_n plots in case of ^{151}Sm$_2$O$_3$, ^{232}Th, ^{209}Bi (n, γ), ^{204,206,207,208}Pb and the fission cross sections of ^{235}U, ^{238}U, and ^{209}Bi nuclei.

6.3 IREN—Facility at Dubna

A new Time of Flight (n_ToF) facility, **I**ntense **Re**sonance **N**eutron Source (IREN) is developed to conduct experiments using gammas and neutrons in the resonance energy region. It is located

Table 6.1 Resonances reported in various experiments. Sample details, energy range of measurement, reported number of resonances, finalized cross sections or the yield, and uncertainty are indicated

Target, mass, size	E_n-range	Number of resonances and energy	Cross-section or kernel value	Reference
^{58}Ni, 2.069 g, $d \times l = 19.91 \times 0.72$ mm^2	0.027–400 keV	51 resonances identified	11 peak CS measured at energies 5–100 keV with ~6% errors	[13] (CERN)
^{63}Ni (impurity of ^{59}Ni)	0.025–200 keV	12 new + 1 ^{59}Ni	Absorption kernels measured	[14] (CERN)
^{151}Sm, 206.4 mg in Sm_2O_3	1 eV–1 MeV	01 + several resonances at E_n = 490–550 eV identified	Maxwellian average, 3100 ± 160 mb	[15] (CERN)
^{186}Os, 1.9999 g	1 eV–1 MeV	Several resonance peaks in 100–200 eV range	A few barns to 100 barns	[16] (CERN)
^{209}Bi, $d \times l = 20 \times 6.1$ mm^2	0.8–23.15 keV	21 resonances identified, 05 cross-section values given at 5, 8, 20, 25 keV and 01 at E_{th}	Resonance parameters with <3% uncertainty and cross sections with ~6% uncertainty	[17] (CERN)
^{232}Th, 2.8037 g, 2 disks dia. = 1.5 cm	3.994–991.452 keV	48, averaged over low and high energy range of corresponding peaks	0.958, 1.281, 1.097, 1.004, 0.912, 0.919, 0.848, 0.817, 0.800, 0.787, 0.761, 0.729, 0.685, 0.613, 0.641, 0.566, 0.545, 0.513, 0.497, 0.468, 0.456, 0.413, 0.365, 0.346, 0.318, 0.275, 0.248, 0.229, 0.220, 0.204, 0.192, 0.172, 0.179, 0.165, 0.158, 0.159, 0.156, 0.147, 0.144, 0.141, 0.140, 0.158, 0.154, 0.164, 0.178, 0.179, 0.156 and 0.135 (all in barns), <3.5% uncertainty	[18] (CERN)
^{238}U, 6.125(2) g, 53.90 × 30.30 mm^2	22 eV–25 keV	Several resonances	Only yield measurements with 2–3% accuracy are reported	[19] (CERN)
^{237}Np (*n*, *f*)	0.2–1000 MeV	Fission cross-sectional curve	CS curve from ~ 0.1 to 2.4 b, CS are ~6% above the evaluated data. Few resonance peaks reported from 37 to 42 eV energy	[21] (CERN)
^{234}U (*n*, *f*)	500–590 eV	Several resonance peaks	Resonance peaks reported from 500 to 590 eV	[21] (CERN)
^{241}Am (32 mg)	0.025 eV–1 MeV	Several resonances	CS not reported	[9] (CERN)
^{94}Mo in 94.025(1) g of ^{nat}Mo	0.01–200 eV	107.2 ± 0.8 eV	CS not reported	[7] (PNF)

(continued)

Table 6.1 (continued)

Target, mass, size	E_n-range	Number of resonances and energy	Cross-section or kernel value	Reference
^{95}Mo in 94.025(1) g of ^{nat}Mo	0.01–200 eV	44.75 ± 0.01, 110.6 ± 1.0, 118.5 ± 1.0, 159.4 ± 0.3 eV	CS not reported	[7] (PNF)
^{96}Mo in 94.025(1) g of ^{nat}Mo	0.01–200 eV	114.7 ± 0.4, 131.4 ± 0.1 eV	CS not reported	[7] (PNF)
^{97}Mo in 94.025(1)g of ^{nat}Mo	0.01–200 eV	71.2 ± 0.1, 78.98 ± 0.59, 109.09 ± 0.84,126.9 ± 1.1 eV	CS not reported	[7] (PNF)
^{98}Mo in 94.025(1) g of ^{nat}Mo	0.01–200 eV	12.2 ± 0.1 eV	CS not reported	[7] (PNF)
^{100}Mo in 94.025(1) g of ^{nat}Mo	0.01–200 eV	97.2 ± 0.7 eV	CS not reported	[7] (PNF)

at the Frank Laboratory for Neutron Physics at JINR from the year 1994 [23]. The facility is to provide pulsed gamma and neutrons with pulse duration of 400 ns and repetition rate to be 150 Hz [24]. The gamma and neutron sources can be used for the studies related to astrophysics, nuclear data, nuclear structure, symmetries and modifications of materials. According to Meshov et al. [25], IREN is a vertically developed facility at JINR Dubna planned for variety of experiments related to ADSS including cross-section measurements. The facility is planned to get an intense neutron source from an electron beam which falls on Tantalum like target and converted into gamma which hits the plutonium-enriched uranium core for conversion and multiplication of neutrons. In a way, neutron source is a subcritical reactor without collection of heat from nuclear energy. Its parameters are given in Table 6.2. Its first phase is already completed in the year 2010 attaining neutron intensity $\sim 10^{11}$ s^{-1}. However, on attaining higher neutron intensity $\sim 10^{15}$ s^{-1}, it will become world's number one facility. Its several characteristic parameters such as mean neutron rate and multiplication factor will be at least ten times of the GELINA facility. With the existing 60 m ToF, cross-section measurements have been made with high accuracy. Experiments such as radiation resistivity of GaAs and Si detectors, analysis of Boron content in Boron-containing ceramics, and analysis of rare elements Pd, Os, Ir, Pt, etc., content of gold-containing ore were conducted among its first experiments. In Fig. 6.6, schematic view of neutron generation is shown without *n*-ToF.

The facility has planned following measurements of elements according to its priority list for the neutron cross-section measurements and covariance [27],

- 19 isotopes of actinides (cross-sectional covariance, nubar covariance) in priority order: ^{235,238}U, ^{239}Pu; ^{237}Np, ^{240,241}Pu, ^{241,242m,243}Am; ^{232}Th, ^{233,234,236}U, ^{238,242}Pu, ^{242,243,244,245}Cm.
- 34 structural, moderator and coolant materials (cross-sectional covariance) in priority order: ^{16}O, ^{23}Na, ^{52}Cr, ^{58}Ni; ^{1}H, ^{12}C, ^{28}Si, ^{90,91,92,94}Zr, ^{206,207,208}Pb, ^{209}Bi, ^{4}He, ^{6,7}Li, ^{9}Be, ^{10}B, ^{15}N, ^{19}F, ^{27}Al, ^{56,57}Fe, ^{155,156,157,158,160}Gd, ^{166,167,168,170}Er.

6.4 TRIGA

TRIGA stands for **T**raining, **R**esearch, **I**sotopes, **G**eneral **A**tomics. Historically, General Atomics (GA) has been pioneer manufacture of fuel rods containing hydrogen in the 1950s. The GA metallurgists perfected making fuel rods of uranium–zirconium hydride (UZrH) as tough as stainless steel and corrosion-resistant fuels. Thus, its use was extended in the 1980s by designing and developing proliferation-resistant fuels

Table 6.2 Planned parameters of the IREN facility [25]

Parameters	Design	Stage 1 (year 2010)
Electron energy (MeV)	200	30
Peak current (A)	1.5	3.0
Beam pulse duration (ns)	200	100
Repetition frequency (Hz)	150	25 ÷ 50
Average beam power (kW)	9.0	0.225 ÷ 0.450
Multiplication target	Pt	W (no multiplication)
Average neutron flux (s^{-1})	1.16×10^{15}	1×10^{11}

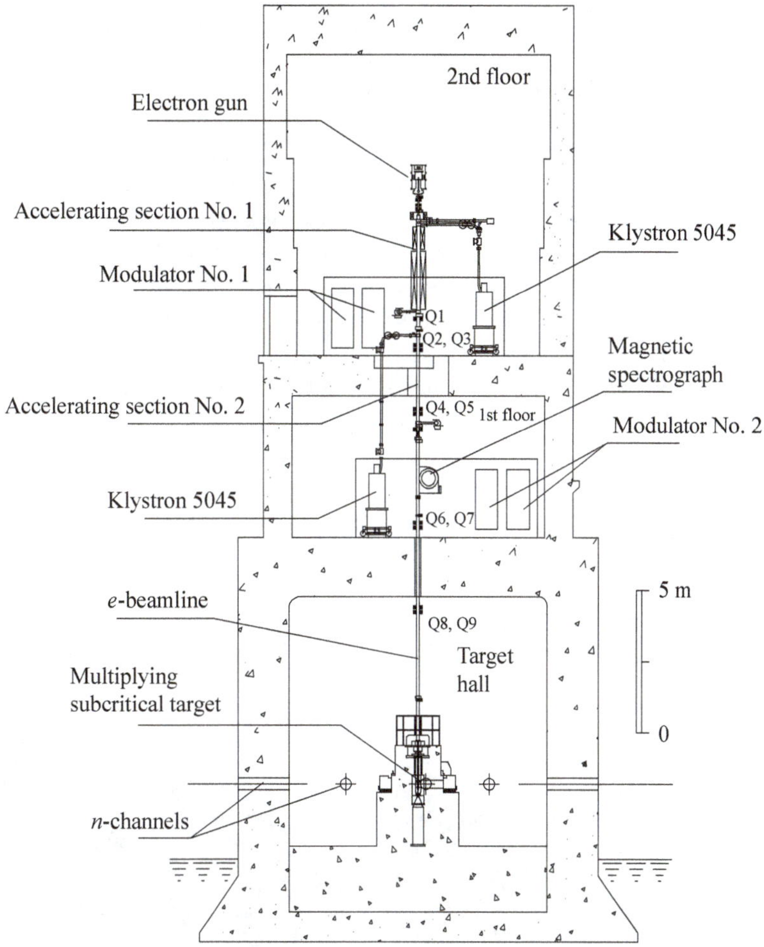

Fig. 6.6 IREN facility to have several neutron channels for experiments [26]

having low uranium content. In a way, TRIGA provides inherent safety than engineered safety by way of external design factors.

General Atomics has installed 66 reactors in universities, institutions, industries, and medical facilities in 24 countries. They have produced low-power reactors ranging in between 0.1 and 16 MW. Some of them are pulsed reactors of high power also. TRIGA International is a joint venture company with CERCA of France which manufactures and sells TRIGA fuel [28] to research reactors. GA has pioneered its scope of research in material production, defense, unmanned spacecrafts, and fusion education outreach, etc. In Fig. 6.7, fuel rods, special customized structures made with SiC materials with required geometries, and neutron reflectors like graphite and Zr_3Si_2 for better fast reactor efficiency are shown. GA has worked for several computational tools such as TRIGLAV [29] for burn-up calculations.

TRIGA performed several innovative experiments for the development of reactor fuels and validate the developed software packages. A 250 kW light water pool-type reactor TRIGA Mark II of Ljubljanawas has been developed by GA [30] using the TRIGLAV calculation package.

Pulsed TRIGA reactor at Mainz has been utilized to develop ultra-cold neutron (UCN) source. The thermal neutrons are cooled down to 6 °K. By putting the converter close to the reactor at $d \sim 39.5$ cm, UCN storage rate ~178,000 UCN per reactor pulse is achieved in the experiment. Maximum density achieved was 550,000 UCN per pulse [31]. As already known, a cold neutron may provide its better decay rate which is demanded in several cosmological studies and even in particle physics. The UCN itself is likely to provide a source for the new neutron spectroscopy in place of scalar optics, and it has high importance over the scalar optics as a neutron is highly penetrating in bulk of a material. In this experiment, UCN density is also observed decreasing with time due to decay.

6.5 KEK Setup

Japan's special attention on research with spallation neutrons started taking shape in the year 1999 [32–34] when Japan Atomic Energy Research Institute (JAERI) and High Energy Accelerator Research Organization (KEK) planned for spallation source of 1 MW proton beam injection with 3 GeV of proton energy and 333 μA of current with pulse duration of 25 Hz in which two 100 ns bunches are injected with 400 ns interval. Beam power ~220 kW is already achieved in February 2017. The spallation target is assumed to be Mercury (Hg). Monte Carlo simulation studies [35] are planned with combined NMTC/JAERI and MCNP. The design is unique from the point of crossing target and moderator system. In fact, it is termed as Target–Moderator–Reflector (TMR) system. Liquid hydrogen is used as the moderator flowing in reverse phase to the flowing Hg target and heavy water (D_2O) as the reflector [34]. Stainless steel SS316 is used as the container, and its related issues of radiation damage have been studied in case of other such facilities. The facility named as 'J-PARC 1 MW pulsed spallation neutron source JSNS' is

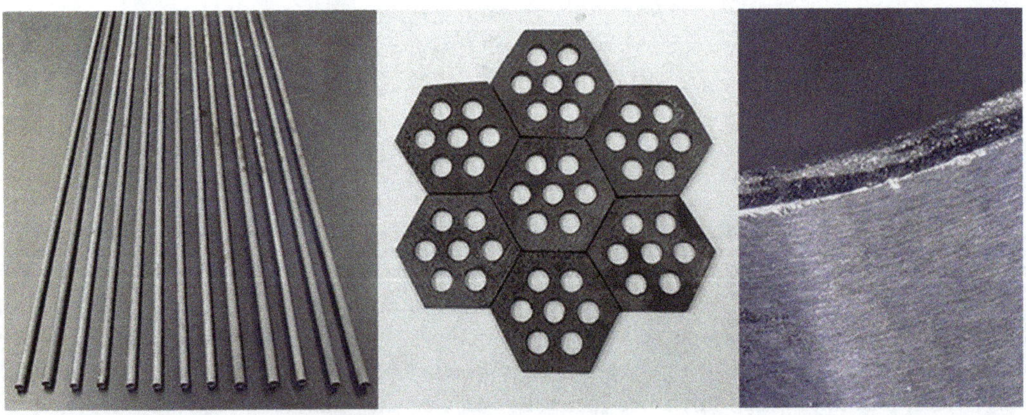

Fig. 6.7 Advance materials as fuel rods (left), structured geometries of special materials like SiC (middle), and neutron reflecting materials such as Zr_3Si_2 (right) [28]

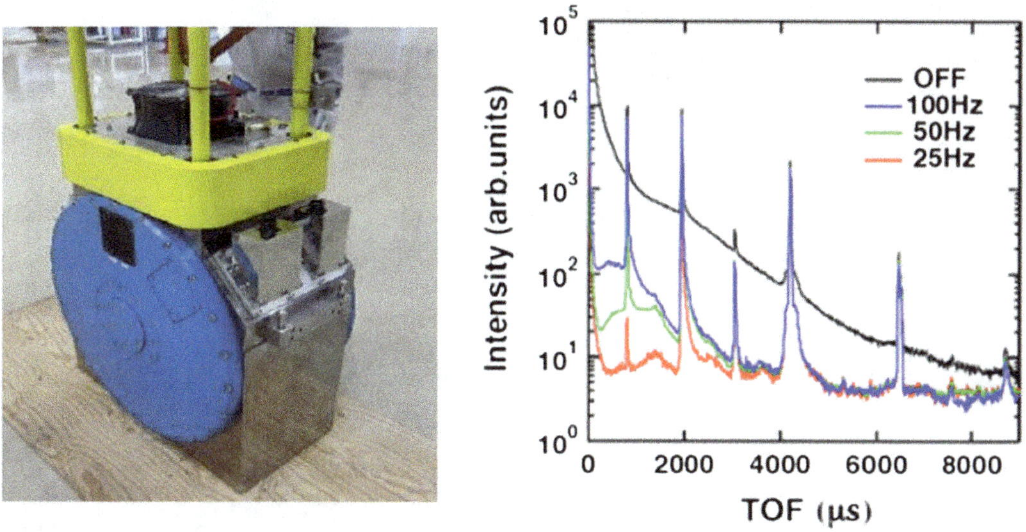

Fig. 6.8 (Left) T0 neutron chopper (right) ToF spectra for operation frequencies $f = 25$, 50, 100, and 'off' modes. Use of a chopper has an advantage of increasing the intensity of the pulse by shortening the path length and cutting down transport losses [36]

successfully launched in May 2008 [36, 37]. Besides, the center caters several developmental activities such as neutron detectors, supermirror devices, ^{3}He neutron spin filters, and choppers with high performance for material science research. With a neutron chopper, finer neutron bunches/burst can be obtained with the change of the chopper frequency as shown in Fig. 6.8.

At the JA-PARC, cold neutron facility is also developed by gradually thermalizing neutrons in heavy water having temperature gradient from 300 °K to less than 80 °K followed by cooling in He-II environment up to 10^{-3} °K [37]. At low temperatures, neutron momentum is transferred by way of phonons. At this stage, UCN are extracted in vacuum up to the cryogenic window placed in a superconducting magnet for the purpose of polarization. Various experiments are conducted with the UCN beam so attained.

In place of a reactor to be a neutron source, attempts are made for a spallation neutron source to be utilized for producing the UCN as shown in Fig. 6.9.

For the neutron scattering experiments, wherever focused neutron beams are required, achromatic focusing optics is important, hence, the 'focusing mirrors.' The purpose of the focusing mirror is to provide narrow area focus of a beam of wide energy range. This is achieved by depositing NiC/Ti on quartz glass. A wide-band neutron beam with wave length $\lambda = 3.64$ Å is focused within 0.25 mm in an attempt made in this direction.

6.6 BFS Setups

According to Russian language, BFS stands for 'Bol'shoy Fizicheskiy Stand' means 'Big Physical Facility' in English language. In the year 2009, Russian federation started a plan of development of Generation IV reactors. Thus, under the ROSTAM 2009, Russia Government decided the following plan to develop next generation nuclear energy technology on the basis of fast neutron reactors with a closed nuclear cycle [38]. Three critical assemblies, BFS-73-1, BFS-75-1, and BFS-76-1A, were created in BFS-1 and BFS-2 facilities along with BFS-109-2A created at IPPE, Obninsk recently in the year 2012.

The BFS-1 had become critical in the year 1961, and it was upgraded recently in the year

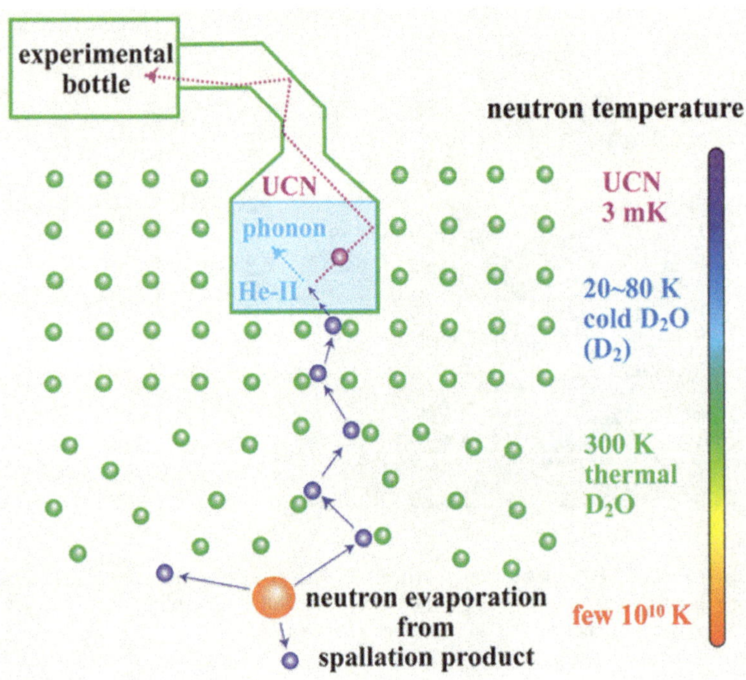

Fig. 6.9 Production of UCN by gradual moderation of the spallation neutrons [36, 37]

2016. It is likely to continue for one more decade. The facility was designed to study full-scale mock-ups of research and fast-power reactors (E_{therm} to 1,000 MW) with different types of fuels including the fertile fuel and coolant (Na, Pb, Pb–Bi, water, gas, and air) and various core and blanket layouts for conducting research for neutronic characteristics of critical assemblies. It is a vertical system with size being $d \times l = 2$ 2.2 m^2 having an experimental opening of diameter ~35 mm. Its critical parameters are given in Table 6.3.

Following experiments have been conducted at the facility,

1. Mock-up of IBR-2, BOR-60, and BN-350, MBIR, SVBR fast Russian reactors as well as foreign reactor mock-ups have been studied.
2. Studies to justify reactor safety of sodium- or lead-cooled fast reactors and VVER-type reactors are conducted.
3. Performed experiments for verification of techniques and specifications of neutron data as well as computer codes for the neutronic characteristics of fast and VVER-type reactors.
4. Experiments on safety justification of the fuel cycle and geological disposal are performed.

BFS-2 is a larger version of BFS-1 critical facility having size $d \times l = 5 \times 3.3$ m^2, about 10,000 fuel tubes carrying fuel in the form of disks. Its power is 1 kW compared to 0.2 kW of BFS-1 facility. Its maximum neutron flux density is an order smaller than BFS-1 and reflector and coolant same as that of BFS-1. Fuel enrichment proportions are similar as that of BFS-1. Power of the reactor facilities has been 3,000 MW than 1,000 MW of the BFS-1.

The setup of BFS complex has been used for several experimental studies under commissioning agreements with the USA, China, Korea, France, Japan, India, and other countries.

6.7 MUSE and YALINA Setups

A comprehensive MUSE program (**Mu**ltiplication with an **E**xternal **S**ource) was started at the MASURCA facility in Cadarache, France [39], where a neutron generator consisting of a deuteron accelerator and a tritium target was coupled to a

Table 6.3 Key technical parameters of the BFS-1 facility

Characteristic	Parameter
Power	0.2 kW
Moderator for simulated light water reactors	Distillate, boric acid solution, polyethylene, graphite
Simulated coolant	Na, Pb, Pb–Bi, water, gas
Reflector	U, UO_2, Pb, Pb–Bi, steel, etc.
Fast neutron flux density, max.	10^{10} cm^{-2} s^{-1}
Core cooling	Natural convection or forced air cooling

subcritical core. Major part of the program was devoted to the investigation of methods for reactivity determination, pulsed neutron source methods, neutron noise methods, and the source jerk method. Some neutron statistics experiments such as fission rate distributions and spectral distributions have also been performed. In the beginning in year 1996, Cf neutron source was used for triggering. Later in the year 1998, generator GENEI 26 (**Ge**nerateeur de **Ne**utrons **I**ntense) was used. Its repeatability rate and wide pulses did not allow precise measurements. Thus, in the MUSE-4, GENEPI-1 (**GE**nérateur de **N**eutrons **P**ulsé **I**ntense), deuteron sharp pulse (1 μs, 50 mA peak current) was used to produce 14 MeV neutron pulses with high repeatability rate. The facility was used for conducting several experiments during 2002–04 and finally dismantled in the year 2007.

Similarly, in the 6th EU framework, European research program for transmutation of high-level nuclear waste by way of an Accelerator-Driven System, IP-EUROTRANS was started. Under this activity, a zero-power subcritical facility, YALINA was installed. It is located at the Joint Institute of Power and Nuclear Research in Sosny outside Minsk, Belarus. Although YALINA does not fulfill the conceptual design of a future ADSS, yet, the neutronics of the subcritical core is an interesting feature as it can be applicable in both fast and thermal systems independent of the type of neutron source. The success of construction of the facility can be assumed as a necessary step toward a full-scale ADSS from the point of understanding the behavior of subcritical cores and the coupling between the main components, the accelerator, the target, and the core.

The facility is constructed to validate and test a possible online reactivity monitoring technique, and its booster is a subcritical fast and thermal core coupled to a neutron generator. As in MUSE-4, the generator in YALINA is an accelerated deuteron ion colliding with Ti target to produce 14 MeV neutrons. The deuteron ion can be in pulsed or continuous mode; hence, the neutron yield to be pulsed or continuous. As an example, in continuous mode, 1.5 mA current of the deuteron projectile can produce $\sim 10^{11}$ neutrons/s.

Both lead zone and the thermal polyethylene zones are shown in Fig. 6.10 with the Ti target in its center. The lead zone is termed as the booster of the primary neutrons produced by Ti target. In the innermost part of the lead booster, UO_2 with 36% enrichment or metallic uranium with 90% enrichment is acceptable as the alternative. Further, the inner fuel configuration is surrounded by 36% enriched UO_2. This is also known as the outer booster. Outside the rectangular configuration as shown in the sectional view, the thermal zone is loaded with UO_2 with 10% enrichment and the polyethylene (C_2H_4) assumingly a better moderator than heavy water. The two zones are separated by a thermal neutron filter or a valve zone of 108 pins of natural uranium and 116 pins of borated carbide, B_4C located in two outside rows of the fast zone. The thermal neutrons migrated from the thermal zone will be absorbed by the boron or ^{nat}U. Thus, the coupling of fast neutrons between the zones is maintained. The B_4C rods inserted in the thermal zone allow change of the reactivity of the system. Positions of extra control rods are also shown in Fig. 6.10. All the zones are enclosed inside the graphite

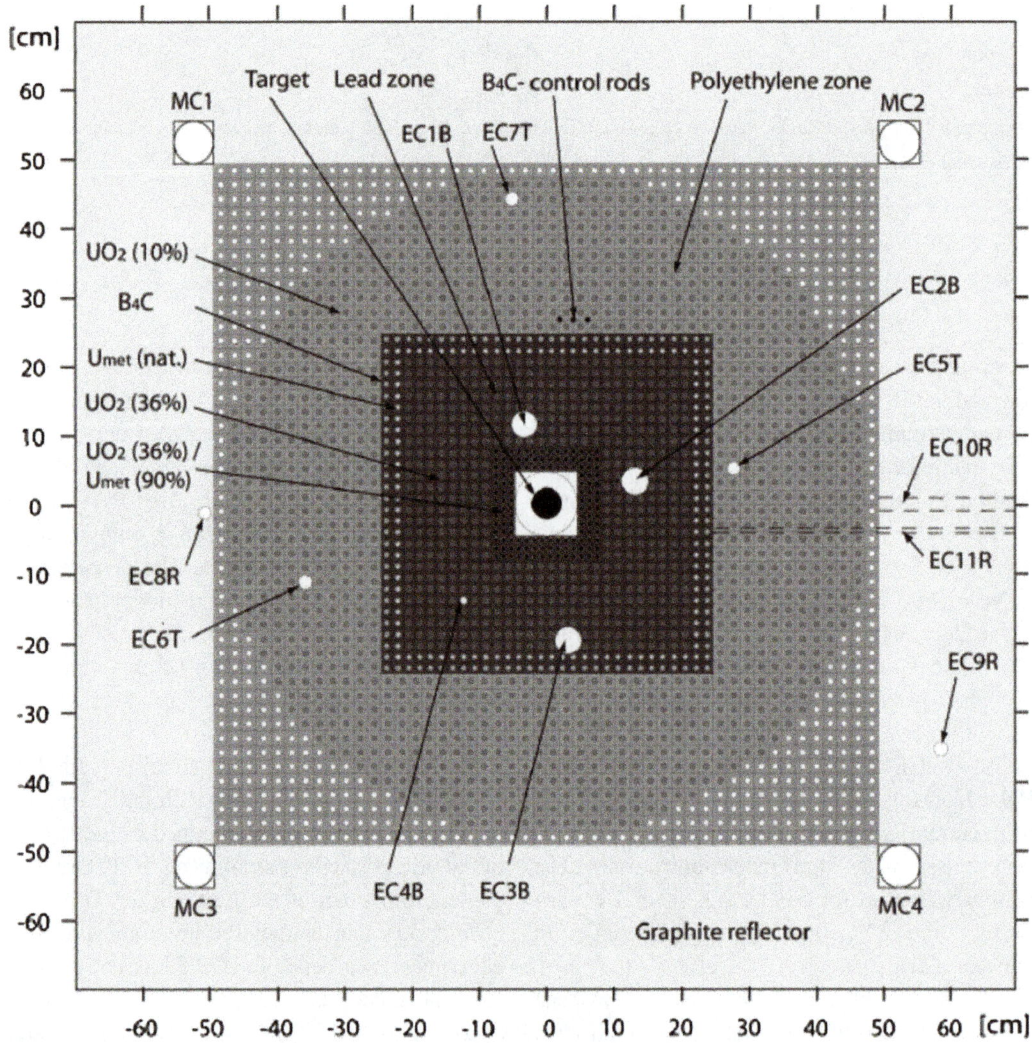

Fig. 6.10 A schematic view of YALINA booster reactor core with possibility of fast-thermal neutron flux zones [40]

reflector. The experimental channels are marked with EC or MC. For details, a reader is advised to follow Berglöf et al. [40].

In Table 6.4, data of criticality, k_{eff} estimated using MCNP code is shown for different loading configurations of the fast and thermal zones. It is understandable that high criticality and small criticality conditions can be achieved by changing configurations of the two zones. Effect of insertion of control rods is also studied. It may be pointed out that except the enrichment details of subcritical core, SC6 all other zones were studied by inserting controlled rods or without control rods.

We understand that changes in a reactor are quantified by the reactivity. A critical reactor has zero reactivity, a subcritical reactor has negative reactivity, and a supercritical has positive reactivity. Reactivity is measured in units as pro cent mille (pcm) and defined as, ρ

$$\rho = 1 - \frac{1}{k_{\text{eff}}} \tag{6.9}$$

Assessment of k_{eff} may be done from MCNP code using its KCODE option. Reactivity monitoring techniques have been described in Refs.

Table 6.4 Enrichment conditions of the booster zones of subcriticality and number of fuel pins, and expected criticality, k_{eff} [40]

Zone	Inner booster		Outer booster	Thermal zone	Expected k_{eff}
Enrichment level	90%	36%	36%	10%	
SC0	132	–	563	1141	0.977
SC3a	–	132	563	1077	0.950
SC3b	–	–	563	1090	0.950
SC6	–	132	563	726	0.850

[40–42] and a according to the references, a neutron or a pulse of neutrons is introduced in the system which will give rise to prompt neutrons, say A_p and later delay neutrons are emitted and they may be called say A_d. An integral view of A_p and A_d measured with the help of detectors installed in the booster system can be used to measure the reactivity.

6.8 GUINEVERE and VENUS-F Setup

GUINEVERE stands for '**G**enerator of **U**ninterrupted **I**ntense **Ne**utron at the lead **V**enus **Re**actor.' It was launched in the year 2006 to operate with VENUS-F reactor at SCK_CEN site in Mol (Belgium). In this setup [43], a reactor is coupled vertically with GENEPI-3C neutron generator to operate in both pulsed and continuous modes. The GENEPI-3C machine is 250 kV deuteron accelerator with a copper target with titanium–tritium (TiT) or titanium–deuterium (TiD) deposits to provide 14 MeV or 2.5 MeV neutrons following T (d, n) ^{4}He or D (d, n) ^{3}He reactions, respectively. The machine is different to MUSE from the point of pulse size as well as being in continuous mode. In its DC mode, the beam can also be used in beam-trip mode. The beam spot size is 20–40 mm diameter, and intensity of produced neutrons would be $\sim 5 \times 10^{10}$ n/s. The VENUS-F core consists of fuel assembly (FA) arranged in a cylindrical geometry ($\sim$80 cm in diameter, 60 cm in height) has fuel $\sim$30% ^{235}U-enriched metallic uranium. In the setup, the core neutrons will be moderated and reflected with solid lead (zero-power experiment) and all experimental details have been discussed by Baylac [44]. In a way, it is a prototype of MYRRHA innovative ADSS. Particularly, GUINEVERE has a vertical coupling with a fast reactor cooled by lead.

Thus, as pointed out above, the GUINEVERE setup is designed to provide a unique experimental setup with continuous beam coupled to fast subcritical assembly for providing results for the following research themes,

(i) validation of the methodology of measuring the subcriticality level,
(ii) investigation and validation of the operational procedures for a future ADSS,
(iii) validation of neutronic codes, and
(iv) safety and licensing issues related to a fast spectrum of an ADSS.

6.9 PURNIMA and KAMINI Experimental Reactors

PURNIMA is abbreviated for **Plu**tonium **R**eactor for **N**eutronic **I**nvestigations in **M**ultiplying **A**ssemblies. PURNIMA-1 was built in 1970s [45] by BARC, Trombay, India, as India's zero-power reactor fueled with plutonium oxide as its first experimental setup. Design of its fuel pin is shown in Fig. 6.11. The fuel pin carries PuO_2 pellets of 1 cm diameter tightly sandwiched between 8-cm-long molybdenum reflectors on both sides. Total length of fuel pin is 41 cm.

Isotopic composition of fuel by weight percentage is ^{239}Pu:^{240}Pu:^{241}Pu::95.75:4:0.25%. A detailed report of the setup is given by Basu et al. [46]. The setup was made critical at criticality = 1 at temperature 25 °C. With the given source intensity $\sim 1 \times 10^7$ n/s, fission rate is estimated to be 3.3×10^{10}/s producing

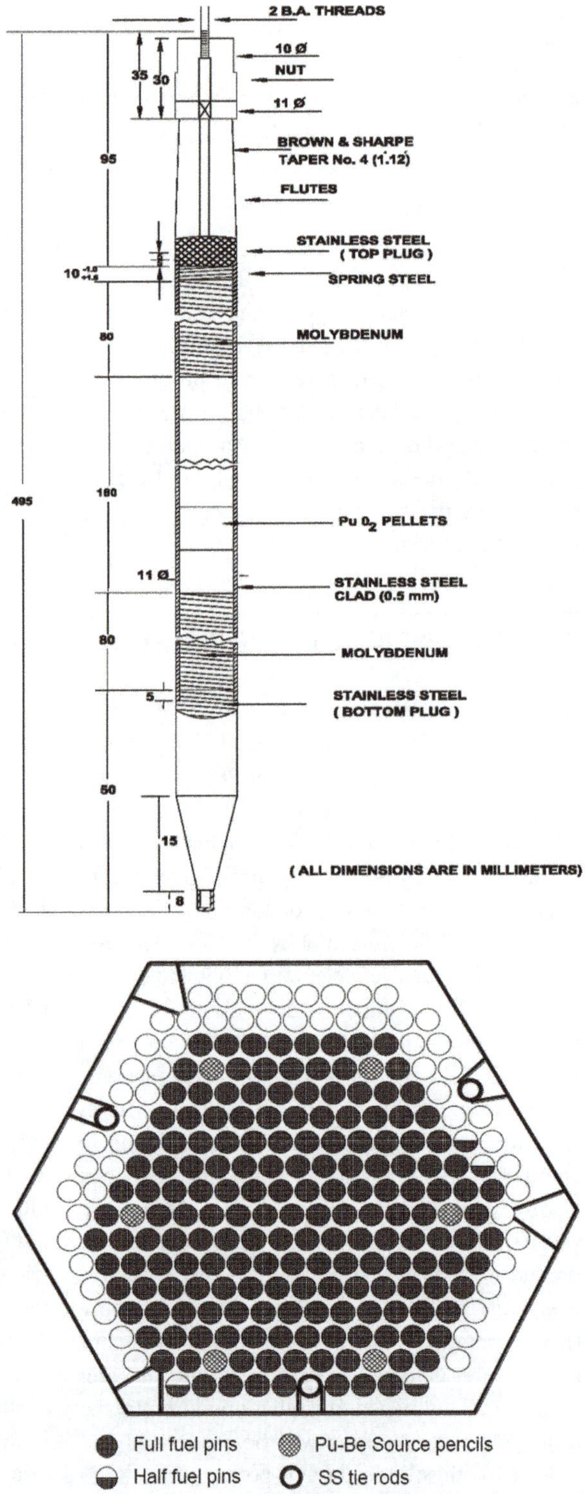

Fig. 6.11 (Up) Design of PuO_2 fuel pin and (below) core configuration of PURNIMA-1 with 175 full and 4 half fuel pin assemblies [45]. Assembly is triggered by Pu–Be neutron source. Hollow pins are dummy

8.2×10^{10} n/s. Due to the fact that source neutrons are at the periphery of the setup and fission neutrons are in the core, then the criticality will be somewhat reduced for the reactor power of 1 W. At 25 °C, criticality is deduced to be 1.00039 ± 0.00441.

PURNIMA-2, a thermal reactor, had reached criticality in May 1984. The reactor was fueled with uranyl nitrate solution (98.1862 weight percent of ^{233}U in $^{233}UO_2(NO_3)_2$) and beryllium oxide (BeO) as reflector. About 400 g of fissile ^{233}U produced in India was burnt in the reactor. This is purely an Indian reactor for the experimental activities like measurement of critical mass as a function of concentration of the solution, reactivity for various safety devices, and measurements of void and temperature coefficients of the reactivity. Details of the design, configuration, other experimental details and evaluations can be seen in Refs. [47–49]. Cross-sectional view through the center of the reactor is shown in Fig. 6.12 [49] with all its components and channels.

The uranium concentration used for the given configuration was 116.6 g/L for a critical mass of 457 g. The reactor was triggered by ^{252}Cf of strength $\sim 10^5$ n/s. Subsequently, the reactor was made critical with uranium concentrations of 103.8, 90.9, 81.5, 70.7, and 60.5 g/L also. It is also planned to be triggered by *d-t* neutron source [50]. Thus, the PURNIMA-2 reactor was decommissioned to produce PURNIMA-3 which became critical in November 1990, and subsequently, it paved way of design of KAMINI reactor.

KAMINI stands for **KA**lpakkam **MINI,** a zero-thermal-power reactor based on ^{233}U fuel (uranium–aluminum alloy) and moderated by light water and BeO being its neutron reflector. It is situated at IGCAR, Kalpakkam, an atomic power research center of Department of Atomic Energy, India. KAMINI became critical in October 1996. Figure 6.13 shows a schematic diagram of the reactor. The size of fuel assembly is $l \times b \times d = 275 \times 66 \times 66$ mm^3 having fuel plates each of size $260 \times 62 \times 2$ mm^3 and wrapped in aluminum frame.

6.10 BRAHMA Subcritical Facility at BARC

Sinha et al. [52] have reported progress of construction of a subcritical assembly developed at BARC, Trombay, with the objective of testing criticality of different fuels and dynamic and static features of neutronics of ADSS. The facility is driven by *d-t*/*d-d* neutron source; hence, coupling of an accelerator may be validated. So far preliminary results of flux and reactivity measurements using pulsed neutron source techniques have been obtained and they are presented in the article by Sinha et al. [52].

6.11 GAMMA Series of Experiments at JINR, Dubna

At JINR, Dubna, a large number of experiments are conducted by producing neutron spectra in collision of accelerated particle with a spallation target. Such neutron spectra are allowed to fall on a sample to produce a nuclear reaction. Several energy neutrons of the spectrum can produce a desired reaction, and the reaction rate is measured using the gamma spectroscopy. Cross section, σ of a reaction can be derived from the reaction rate, $R = \sigma \times \Phi$, where Φ is the incident particle flux expressed in units as n/cm^2/s and σ in cm^2. In case of particle flux corresponds to a given energy spectrum in place of unique energy, cross section is known as an spectrum average cross section (sp. av. cs.) [53] and given by

$$\sigma_{\text{sp.av}} = \Sigma\sigma_i\phi_i/\Sigma\phi_i \tag{6.10}$$

and the flat cross section σ_i of the *i*th energy bin in between E and $E + \mathrm{d}E$ is given by,

$$\sigma_i = \int\sigma(E)\mathrm{d}E/\int\mathrm{d}E \tag{6.11}$$

For the calculation of $\sigma_{\text{sp.av}}$ from Eq. (6.10), data of cross section, σ_i corresponding to neutron energy can be taken from the ENDF files if it is available at the given energy. For discretely available data, first flat cross sections are

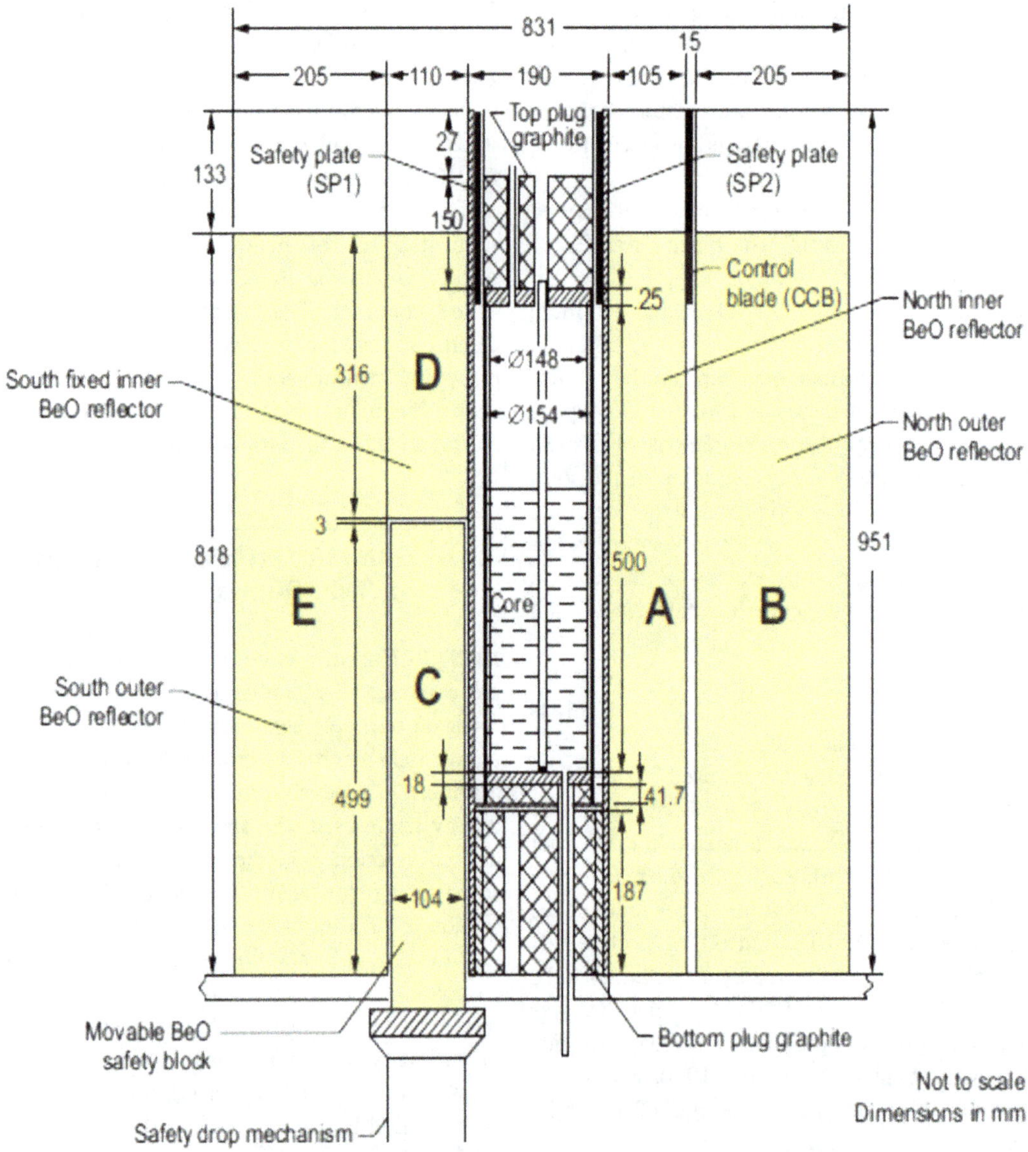

Fig. 6.12 A vertical cross-sectional view of the reactor through its center [49]. All dimensions are in mm

calculated by dividing the entire energy range into a number of groups as shown by Eq. (6.11).

As an example flat cross sections obtained from the evaluated data of ENDF/B VI.8 data file of Th (n, γ) reaction are plotted as function of neutron energy in Fig. 6.14. Thus, for a given neutron energy, 'spectrum average cross section,' $\sigma_{\text{sp.av}}$ are calculated [55] from the flat cross sections obtained from standard PREPRO software. A series of GAMMA experiments have been performed using different neutron energy spectra by varying proton energy or by changing medium in between the spallation target and the sample which moderates the spallation neutron spectrum.

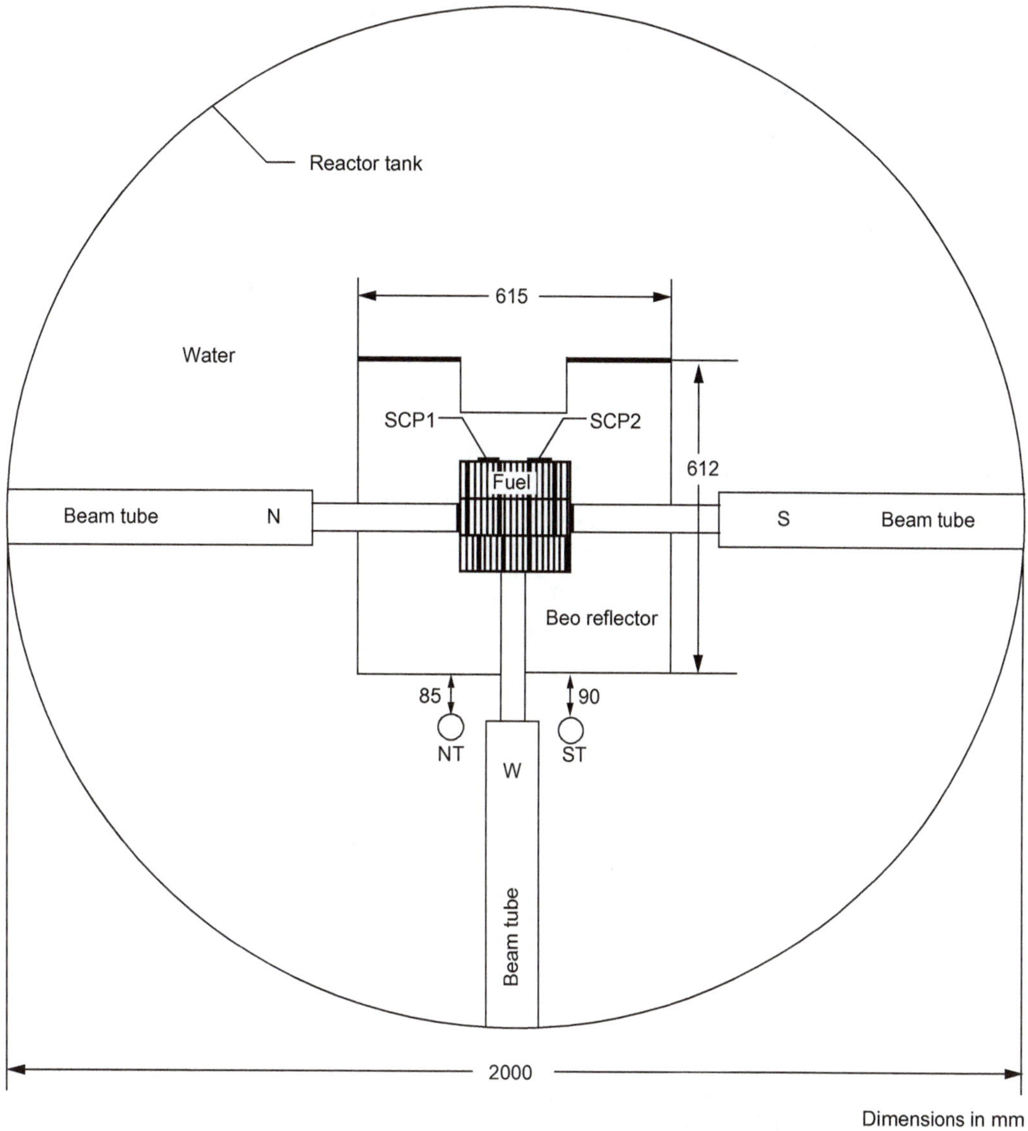

Fig. 6.13 KAMINI fuel and reflector assembly shown with three beam channels [51]

6.11.1 GAMMA-2

It is a setup where spallation neutron spectrum is produced from the impact of 1 GeV proton with lead target of dimension, $2R \times L = 8 \times 20\ \text{cm}^2$. The lead target is covered with 6-cm-thick layer of paraffin $(CH_2)_n$ moderator. As calculated, 1 GeV proton will not lose its entire energy

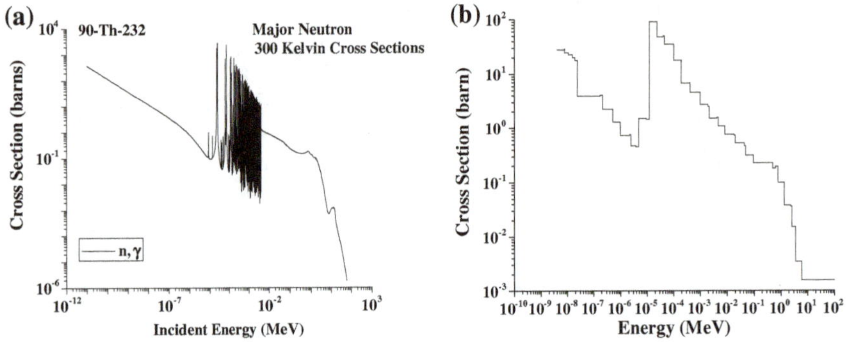

Fig. 6.14 (**a**) ENDF/B-VI.8 cross section data of ^{232}Th (n, γ) reaction (**b**) 'flat' cross section plot for the same reaction after processing by the PREPRO software for the energy intervals considered in Dubna CASCADE code ver. 2004 [54].

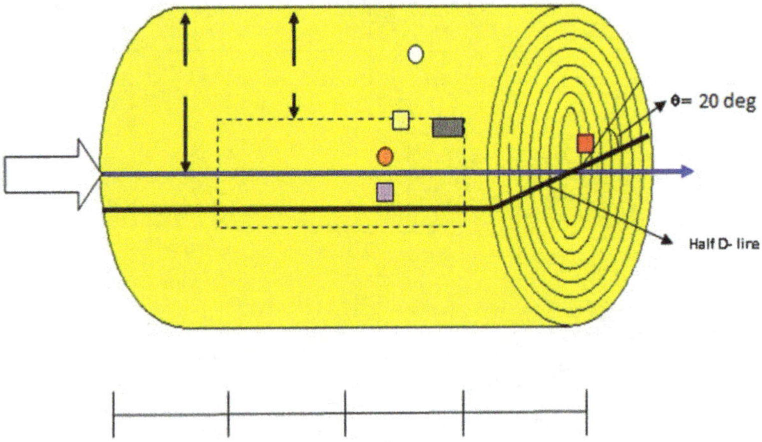

Fig. 6.15 GAMMA-2 setup showing lead spallation target inside the paraffin layer with proton beam impact on the left front. Activation sample of Th placed on lead and others on the paraffin surface is shown [55]

Positions of samples are shown in the following,

→ ^{232}Th-sample (size 3.2 × 2.5 0.00015 cm^3) placed at the interface of paraffin and Pb at $z = 25$,

→ ^{197}Au-sample placed at $z \sim 20$ cm on the outer surface of the paraffin,

→ ^{209}Bi-sample placed at $z \sim 20$ cm on the outer surface of the paraffin,

→ ^{115}In-sample placed at $z \sim 20$ cm on the outer surface of the paraffin,

→ ^{59}Co-sample placed at $z \sim 20$ cm on the outer surface of the paraffin,

→ ^{181}Ta-sample (1.6 × 1.65 × 0.0431 cm^3) placed on the outer paraffin circular face of upper paraffin dee at $r = 4$ cm, $z = 32$ cm and $\theta = 20°$.

by way of ionization loss in passing through 20 cm length of lead including layer of paraffin in the situation that no nuclear collision takes place. In Fig. 6.15, lead + paraffin setup of the target is shown with proton beam striking at the left-side end of the lead target. In the lower part of the figure, a realistic experimental setup with the positions of different activation samples is shown.

The experiments are conducted to estimate neutron fluxes in different positions of the setup.

Incident proton beam distribution on the left end of the lead target at different radial distances is estimated from the Al-monitor detectors using the ^{27}Al (p, $3pn$) ^{24}Na reaction. The Al monitor is placed at −60 cm distance from the left end of the Pb target. Using γ-spectrometry of the activated Al foils, activity of the ^{24}Na yield is estimated to conclude intensity of beam which comes out to be 1.09E + 09 protons/s for the irradiation time of 5.4836 h [53]. Using the CASCADE code ver.04 [54], neutron flux in (a) longitudinal position of ^{232}Th, (b) moderated neutrons in positions of four activation samples on cylindrical surface, and (c) moderated neutron on the forward face at the position of ^{181}Ta sample are simulated for the incident 1 GeV proton beam and given in Fig. 6.16a–c.

Using the cross-section data of (n, γ), (n, $2n$), and (n, $4n$) reactions from the ENDF VI.8 data files, neutron flux is deduced from the measured reaction rates. Details of the experiment and reaction rates are given in Ref. [55], and some of the results are summarized in Table 6.5. It may be inferred that

(i) Incident neutron flux on a sample in its given position is closely similar when estimated using the two reactions, e.g., (n, γ) and (n, $2n$) in case of ^{232}Th.
(ii) Neutron flux is found to be in decreasing order when compared at axial position of four activation samples and at the forward face of the paraffin cylinder.

At the end face of the setup where ^{181}Ta sample is placed, several remnant beam protons also produce reaction ^{181}Ta (p, pn) ^{180}Ta. Correction in the neutron flux corresponding to the (n, pn) reaction for producing ^{180}Ta is estimated [53]. Based on the neutron spectrum estimated as above, Adam et al. [55] have deduced 'spectrum average cross sections' for a large number of reactions including high-order (n, xn) reactions with x = 3–9 and several other reactions. They are given in Table 6.6.

6.11.2 GAMMA-3

The experimental setup GAMMA-3 provides a neutron flux produced by collision of 2.33 GeV deuteron beam with the lead target (size $d \times l$ = 8 × 60 cm^2) which is subsequently moderated by the big block of graphite [56, 57]. The experiment was conducted at JINR Nuclotron for comparing with the results of TARC experiment [58] where n-flux was moderated by a large block of lead. The setup is different to the GAMMA-2 from the point of beam and the moderator.

GAMMA-3 experimental setup is shown in Fig. 6.17 where the graphite moderator of size 1.1 × 1.1 × 0.6 m^3 is made of 25 blocks of different sizes and the setup carries several experimental holes like 'a', 'b,' and 'c' particularly marked on the setup. In the middle, spallation target is placed and marked with letter, T. Sizes of holes 'a', 'b,' and 'c' are $d \times l$ = 14.6 × 29.6 cm^2, 8.8 × 36.3 cm^2, and 15.4 × 34.1 cm^2, respectively. In Fig. 6.17, cylindrical structure carrying experimental pockets for placing samples are shown and the similar cylinders are placed inside the holes 'a', 'b,' and 'c'. Hole 'a' is closest to the target, T. Geometrical details of the setup can be seen in Ref. [56, 57]. Samples of ^{nat}U were placed in two holes 'a' and 'b' only. Uranium and thorium samples were irradiated in the form of sandwiches of three very nearly identical foils as Th–Th–Th and U–U–U, and this arrangement has an advantage of accounting for the recoiled residual nuclei produced in the middle foil. Diameter of each foil is 15 mm, and mass of middle U and Th foils are 0.1723 and 0.0931 g,

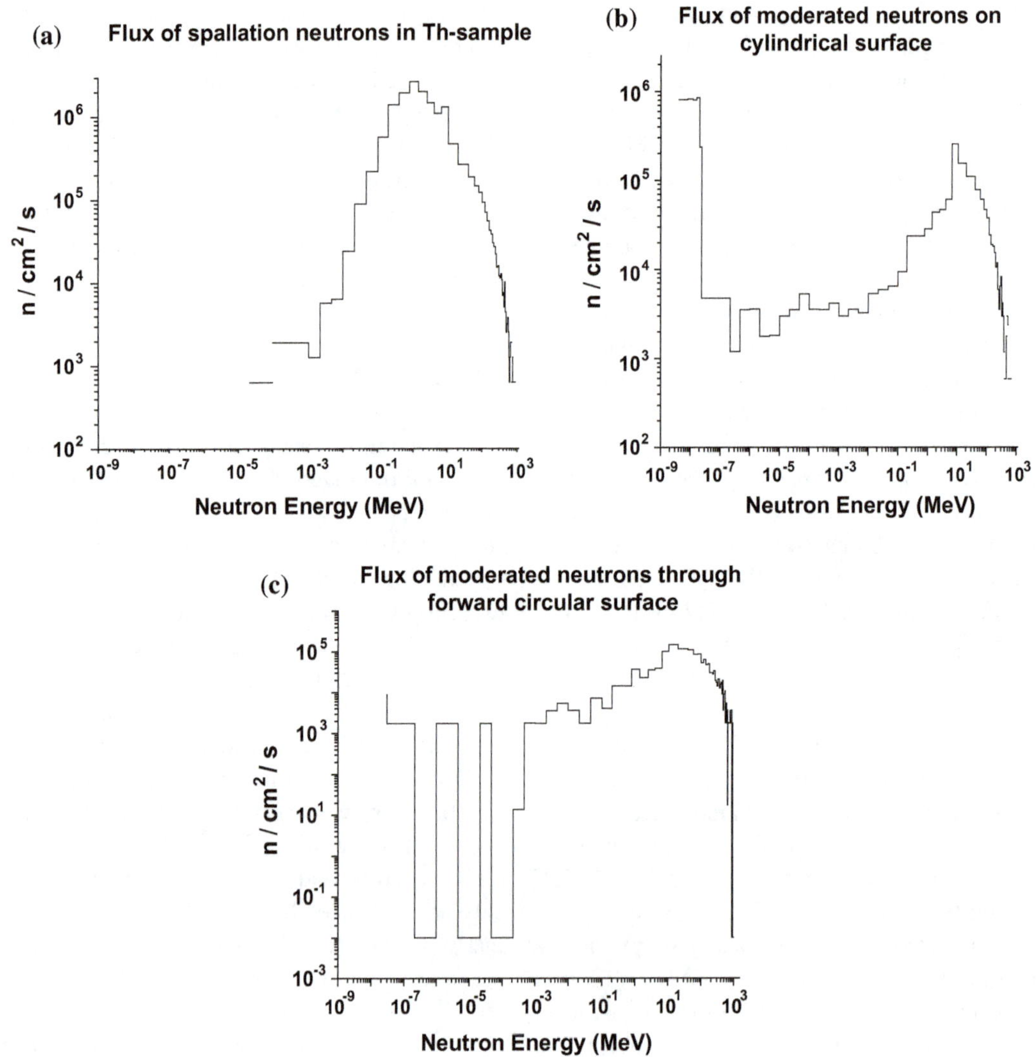

Fig. 6.16 Simulated flux of (**a**) spallation neutrons on ^{232}Th sample (**b**) moderated neurons on paraffin surface at the ring position of four activation samples and (**c**) moderated neutrons on the outgoing forward face at the position of ^{181}Ta sample [54]

respectively. The total masses of the sideward ^{nat}U and ^{232}Th foils are 0.334 and 0.1763 g, respectively.

In the year 2007, spallation Pb target of the setup was irradiated with 2.33 GeV deuteron beam for 25 h 17 m and intensity of the beam was monitored by the reaction ^{27}Al $(d, 3p2n)$ ^{24}Na with the Al foils of thickness 6.95 mg/cm^2 and diameter 20 cm which was installed at a distance of 3.1 m from the center of the lead spallation target in between the beam pipe and the spallation target. From the gamma spectrometry of Al foil, integral number of deuterons reaching the Pb target is estimated to be $N = (1.704 \pm 0.103) \times 10^{13}$.

Monte Carlo code MCNPX v 2.6.C was used [59] for simulation of (i) production of neutrons in collision of 2.33 GeV deuteron on the lead target and (ii) transport of neutrons through the graphite setup. As mentioned earlier that the

Table 6.5 Neutron flux deduced from the measured reaction rates [55] and average cross sections from the ENDF VI.8 compared with the values obtained from the CASCADE code [54]

S. No.	Surface, sample	Nuclear reaction	<σ>	$\Phi = R/\sigma$ (n/cm²/s)	CASCADE code (n/cm²/s) [54]
1	Interface position, ^{232}Th	^{232}Th (*n*, γ) ^{233}Th	64.8 mb	1.50 ± 0.09 × 10^{7}	1.46 × 10^{7}
		^{232}Th (*n*, 2*n*) ^{231}Th	0.253 b	1.44 ± 0.33 × 10^{7}	
2	Outer cylindrical surface of paraffin, ^{197}Au	^{197}Au (*n*, 2*n*) ^{196}Au	0.12 b	0.746 ± 0.023 × 10^{7}	0.64 × 10^{7}
		^{197}Au (*n*, 4*n*) ^{194}Au	0.055 b	0.577 ± 0.022 × 10^{7}	
3	End circular face, ^{181}Ta	^{181}Ta (*n*, γ) ^{182}Ta	1.13 b	1.24 ± 0.29) × 10^{6}	1.43 × 10^{6}
		^{181}Ta (*n*, 2*n*) ^{180}Ta	0.612 b	1.37 ± .06) × 10^{6}	

Table 6.6 Spectrum cross sections ⟨σ⟩ of different reactions deduced from the measurement of reaction rates, *R* [55]

Reaction	E_γ (keV)	E_{th} (MeV)	⟨σ⟩
^{197}Au (*n*, γ) ^{198}Au	411.80	0.0	26.9 ±0 .67 b
^{197}Au (*n*, α) ^{194}Ir	328.40	~0.0	7.72 ± 0.05 mb
^{197}Au (*n*, *p* + 6*n*) ^{191}Pt	538.86	43.31	241 ± 3.7 mb
^{197}Au (*n*, 6*n*) ^{192}Au	316.50	38.92	33.67 ± 1.1 mb
^{197}Au (*n*, 7*n*) ^{191}Au	586.45	45.99	29.63 ± 2.0 mb
^{197}Au (*n*, 8*n*) ^{190}Au	295.80	55.04	08.83 ± 0.14 mb
^{59}Co (*n*, α) ^{56}Mn	846.77	0.0	3.46 ± 0.16 mb
^{59}Co (*n*, 2*n*) ^{58}Co	810.81	10.63	98.0 ± 2.69 mb
^{59}Co (*n*, 4*n*) ^{56}Co	846.77	30.92	1.70 ± 0.12 mb
^{59}Co (*n*, 5*n*) ^{55}Co	931.50	41.18	0.33 ± 0.06 mb
^{209}Bi (*n*, 6*n*) ^{204}Bi	899.17	38.13	19.72 ± 0.36 mb
^{209}Bi (*n*, 7*n*) ^{203}Bi	820.33	45.36	17.05 ± 0.31 mb
^{209}Bi (*n*, 8*n*) ^{202}Bi	960.66	54.29	11.82 ± 0.21 mb
^{209}Bi (*n*, 9*n*) ^{201}Bi	629.14	61.71	10.27 ± 0.21 mb
^{181}Ta (*n*, *pn*) ^{180m}Hf	332.28	5.98	0.53 ± 0.004 b
^{181}Ta (*n*, 4*n*) ^{178m1}Ta	426.38	22.27	0.304 ± 0.013 b
^{181}Ta (*n*, 5*n*) ^{177}Ta	112.95	29.17	0.130 ± 0.006 b

^{232}Th samples were irradiated in all the three holes, namely '*a*', '*b*,' and '*c*' while the samples of ^{nat}U were irradiated in holes '*a*' and '*b*.' All the details of their positions as shown in Fig. 6.17 are included in the simulations. The results of simulated neutron flux falling on the two samples are given in Fig. 6.18. It can be seen that there are humps of neutron fluxes at both low and high energies. Spectrum at low energy is pronounced because of the fact that a large number of neutrons are moderated in the graphite and fall in the thermal region. It may also be noted that at all energies, neutron fluence decreases from hole '*a*' to hole '*c*' because the

Fig. 6.17 (Up left) Graphite moderator block with spallation target at the middle and carrying several experimental holes (up right) packing cylinder having sample pockets and (below) positions of the samples on a cylinder fitted in a hole. ^{nat}U is placed at −9° and ^{232}Th at +9° from the center of the front face of the cylinder in the back of the circles [56]

distance of holes '*a*' from the center of the Pb target is smaller than that of the hole '*c*.'

In Table 6.7, results of reaction rate, *R*, both experimental and calculated [56], are given for the three experimental holes *a, b,* and *c*.

After a complete analysis of ^{nat}U sample irradiated in holes '*a*' and '*b*,' 11 fission products, namely ^{85m}Kr, ^{93}Y, ^{99}Mo, ^{103}Ru, ^{105}Rh, ^{131}I, ^{132}Te, ^{133}I, ^{140}Ba, ^{141}Ce, and ^{143}Ce were observed. Also, ^{239}Np is observed as a product of ^{238}U (n, γ) reaction and ^{237}U as a product of ^{238}U $(n, 2n)$ reaction occurred in the ^{nat}U sample. Similarly, in case of ^{232}Th sample irradiated in the hole '*a*,' six fission products, namely ^{85m}Kr, ^{99}Mo, ^{131}I, ^{133}Xe, ^{135}Xe, and ^{141}Ce are observed. Also, ^{233}Pa product corresponding to ^{232}Th (n, γ) reaction and ^{231}Th product corresponding to ^{232}Th $(n, 2n)$ reaction are observed. In case of ^{232}Th in the hole '*b*' only four fission products ^{85m}Kr, ^{99}Mo, ^{131}I, and ^{141}Ce are observed along with ^{233}Pa and ^{231}Th. Lastly, in case of the hole '*c*,' only ^{99}Mo is observed as a fission product and ^{233}Pa as a product of ^{232}Th (n, γ) reaction. This shows that as we go farther from the spallation source more and more, moderated neutrons have lesser probability of fission reaction to happen with ^{232}Th, partly due to lesser neutron energy and lesser available amount of neutron fluence. Similar inference was also drawn with the help of data of reaction rate, *R* given in Table 6.7.

Using the data of reaction rate, $R\,(A_r, Z_r)$ of (A_r, Z_r) product, transmutation power of a system as derived in Ref. [4] can be written as follows

$$P(A_r, Z_r) = R(A_r, Z_r) \cdot N_{\mathrm{d}} \frac{A_r}{A_{\mathrm{t}}} t_{\mathrm{irr}} \qquad (6.12)$$

Here, $N_{\mathrm{d}}.t_{\mathrm{irr}} = N_{\mathrm{D}}$ (an integral number of deuterons). On normalizing to 10^9 beam particles, one can write

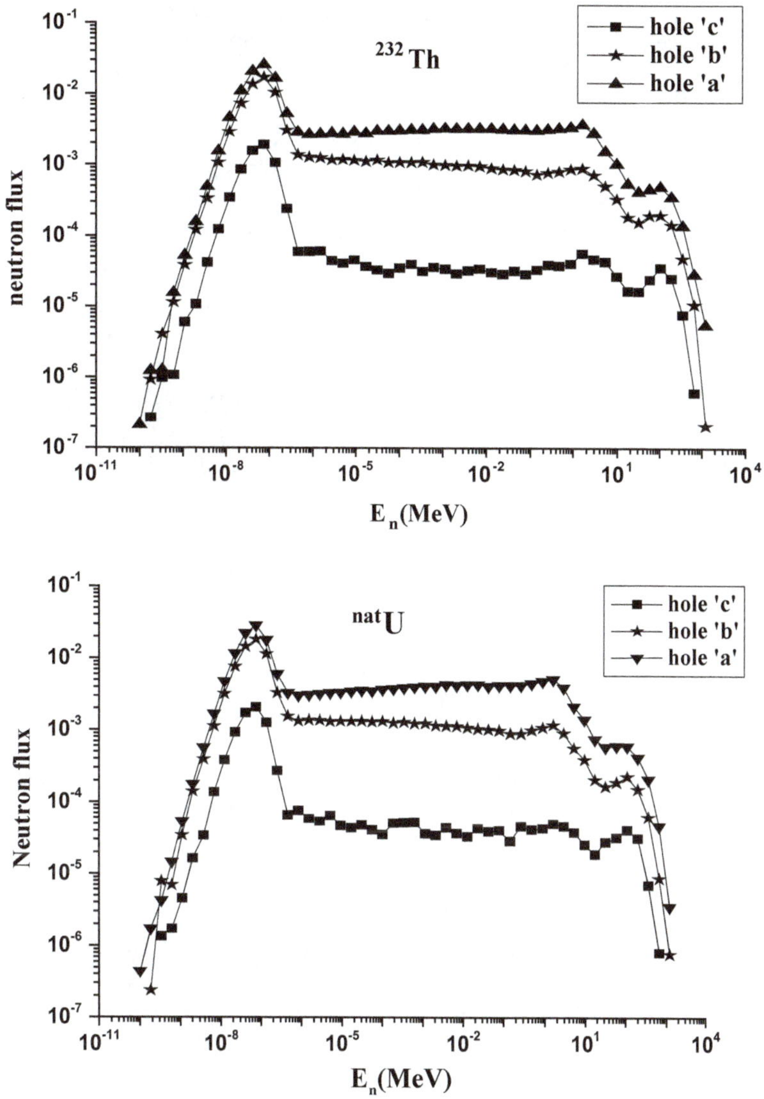

Fig. 6.18 (Up) Simulated neutron flux (n/cm^2/deuteron) at the positions of ^{232}Th sample (on left) and ^{nat}U sample (below) in holes '*a*', '*b*,' and '*c*' [56]

$$P_{\mathrm{norm}}(A_r, Z_r) = 10^9 \frac{P(A_r, Z_r)}{N_{\mathrm{D}}} \tag{6.13}$$

where N_d is the deuteron beam intensity, t_{irr} is the irradiation time and A_t is the target mass number. In Table 6.8, normalized transmutation power P_{norm} in GAMMA-3 experiment is compared for the two other experimental setups, viz. Transmutation by Adiabatic Resonance Crossing (TARC) [58] and $E + T$ [60].

Following observations can be made from the data of Table 6.8

1. Transmutation power in case of (n, γ) reaction for ^{232}Th and ^{nat}U is estimated independently for the two holes '*a*' and '*b*,' and it is found comparable. When we compare with the values obtained in case of different $E + T$ setups [60–62] and TARC assembly [58], then it may be inferred that P_{norm} is about an order of magnitude higher in graphite assembly than other $E + T$ and TARC assemblies.
2. On comparing P_{norm} $(n, 2n)$ for ^{232}Th in case of the graphite setup and the $E + T$ assembly, it may be pointed out that transmutation

Table 6.7 Calculated and experimental values of reaction rates of (n, γ), (n, f), and $(n, 2n)$ reactions for both ^{232}Th and ^{nat}U samples irradiated in the three positions *a, b,* and *c* inside the graphite moderator for the neutron fluencies in the three positions [56]. Values shown in the brackets correspond to the errors, e.g., 3.31(15) = 3.31 ± .15 and 0.49 (1) = 0.49 ± 0.01

Hole position	Reaction	^{232}Th			^{nat}U		
		^{232}Th (n, γ) ^{233}Pa	(n, f)	^{232}Th $(n, 2n)$ ^{231}Th	^{238}U (n, γ) ^{239}Np	^{nat}U (n, f)	^{238}U $(n, 2n)$ ^{237}U
a	$R_{expt.}$ '*a*'	3.31(15)*E*−25	1.06(21)*E*−26	8.9(11)*E*−27	3.38(11)*E*−25	2.42(33)*E*−25	2.83(19)*E*−27
	$R_{cal.}$ '*a*'	7.83*E*−25	5.09*E*−27	3.27*E*−27	1.63*E*−24	2.54*E*−25	2.73*E*−27
	$R_{expt.}/R_{cal.}$	0.423(19)	2.08(41)	2.72(34)	0.207(7)	0.95(13)	1.04(7)
b	$R_{expt.}$ '*b*'	1.96(5)*E*−25	2.55(69)*E*−27	1.06(22)*E*−27	2.31(8)*E*−25	1.57(24)*E*−25	7.47(83)*E*−28
	$R_{cal.}$ '*b*'	4.04*E*−25	1.78*E*−27	7.58*E*−28	6.07*E*−25	1.56*E*−25	7.63*E*−28
	$R_{expt.}/R_{cal.}$	0.49(1)	1.43(39)	1.39(29)	0.38(1)	1.01(15)	0.98(10)
c	$R_{expt.}$ '*c*'	3.38(16)*E*−26	3.16(88)*E*−28	–	–	–	–
	$R_{cal.}$ '*c*'	3.39*E*−26	2.13*E*−28	6.84*E*−29	3.73*E*−26	1.79*E*−26	6.61*E*−29
	$R_{expt.}/R_{cal.}$	0.997(5)	1.48(41)	–	–	–	–

Table 6.8 Comparison of normalized transmutation power, P_{norm} of three assemblies, namely graphite-lead target of GAMMA-3, *E* + *T* [60], and TARC [58]. Distances of the samples from the center of their respective assemblies are given as '*d*'

Assembly	Graphite [56]			*E* + *T* [60]	TARC [58]	
	Hole *a*	Hole *b*	Hole *c*			
Distance '*d*' (Th)	*d* ~ 24 cm	*d* ~ 34 cm	*d* ~ 61 cm	*d* ~ 13 cm	Z = 22.5 cm, X = 122 cm	Z = 7.5 cm, X = 150 cm
^{232}Th (n, γ) ^{233}Pa	3.32(15)*E*−16	1.97(5)*E*−16	3.39(16)*E*−17	3.09(13)*E*−17	3.8(3)*E*−17	1.0(2)*E*−17
^{232}Th $(n, 2n)$ ^{231}Th	8.9(11)*E*−18	1.06(22)*E*−18	–	1.59(16)*E*−18	–	–
Distance '*d*' (U)	*d* ~ 19 cm	*d* ~ 31 cm	*d* ~ 58 cm	*d* ~ 13 cm	*d* = 94 cm, Z = −22.5 cm	*d* = 107 cm, Z = −22.5 cm
^{238}U $(n. \gamma)$ ^{239}Np	3.39(11)*E*−16	2.32(8)*E*−16	–	2.87(9)*E*−17	7.7(2)*E*−17(hole 7)	1.1(3)*E*−17(hole 6)
^{238}U $(n, 2n)$ ^{237}U	2.82(19)*E*−18	7.44(83)*E*−19	–	–	–	–

power is about seven times higher in the graphite setup than the $E + T$ assembly.

3. The above study is highly innovative, and it may have impact on the design of an ADSS for transmutation of LLNW.

6.12 Energy + Transmutation Experimental Setup

In the year 1999, JINR started 'Energy + Transmutation' program of research with the objective of 'investigations of physical aspect of electronuclear energy generation and transmutation of radioactive waste of the reactors using the high energy beams of the Nuclotron, JINR (Dubna)' [62–64]. Several experiments have been conducted using the proton beams of energies ranging from 0.7 to 3.7 GeV under the program. The experiments were focused on general aspects of energy generation, neutron multiplication, neutron spectra, neutron induced transmutation of long-lived minor actinides (specially ^{237}Np and ^{241}Am), fission product (^{129}I), and plutonium isotopes (^{238}Pu, ^{239}Pu). A review of the experiments has been published by Adam et al. [65] along with several other details of experiments. In the year 2005, the $E + T$ experimental setup was irradiated with 2.52 GeV deuteron beam to compare its results with earlier experiments with proton beams. Soon after it, in December 2006 the setup was irradiated with the 1.6 GeV deuteron beam [60]. In Fig. 6.19, basic structure of the $E + T$ setup is shown.

The whole assembly was covered with shielding of size having outer dimensions $\Delta X \times \Delta Y \times \Delta Z = 100 \times 106 \times 111$ cm^3 as shown in Fig. 6.20.

Al foils are used as deuteron beam monitors and the beam is focused on the center of the lead target. Also, special efforts are made to detect centroid of the beam using the track detectors. In Fig. 6.21 (left), results of experimental beam distribution using Al foils and in Fig. 6.21 (right) results of track detectors are presented [60]. In the experiment, integral number of beam hitting the spallation lead target $N_D \sim 1.93(25) \times 10^{13}$.

In simulations using the code MCNPX, the real coordinates of beam-hits, that were obtained from the track detectors, on the target were used. It is also important to mention that beam not only loses slight alignment with the beam axis but also may be asymmetric in time. Being asymmetric, it has definite effect on production and detection of nuclear products and beam correction become essential in calculations of reaction rates. In Fig. 6.22, beam profile of 1.6 GeV deuteron beam hitting the target of the experiment obtained from the accelerator scientists is given for an example.

Using the CASCADE code ver. 04, neutron flux in the position of the RA samples put on the top of the second section of $E + T$ assembly has been simulated [64] in the situation of total shielding as shown in Fig. 6.20. The input files of simulation by the CASCADE code and other details of the $E + T$ setup can be seen in Ref. [53]. In Fig. 6.23, simulated neutron flux of the setup irradiating the RA samples calculated from the CASCADE code and MCNPX are shown.

Looking at the details of the data in Fig. 6.23 in the neutron energy region 0.02–1000 MeV range, the two codes give comparable total flux and in the energy range smaller than 0.01 MeV, flux from the MCNX is nearly an order of magnitude higher than CASCADE. According to MCNPX, total n-flux is 2.17×10^7 n/cm^2/s in good agreement with 2.31×10^7 n/cm^2/s obtained from the CASCADE code. It may be pointed out that the difference in distribution has been reduced in later version of the CASCADE code published as MONC [66].

In the $E + T$ setup, neutron irradiation of highly pure samples of ^{129}I, ^{127}I, and ^{237}Np was done also using the 2.52 GeV deuteron beam, putting samples on the top of the second section. Some of its results are shown in Table 6.9 in case of ^{237}Np [53].

A comparison of transmutation power for the (n,f) reactions of ^{237}Np and ^{239}Pu is compiled [67]

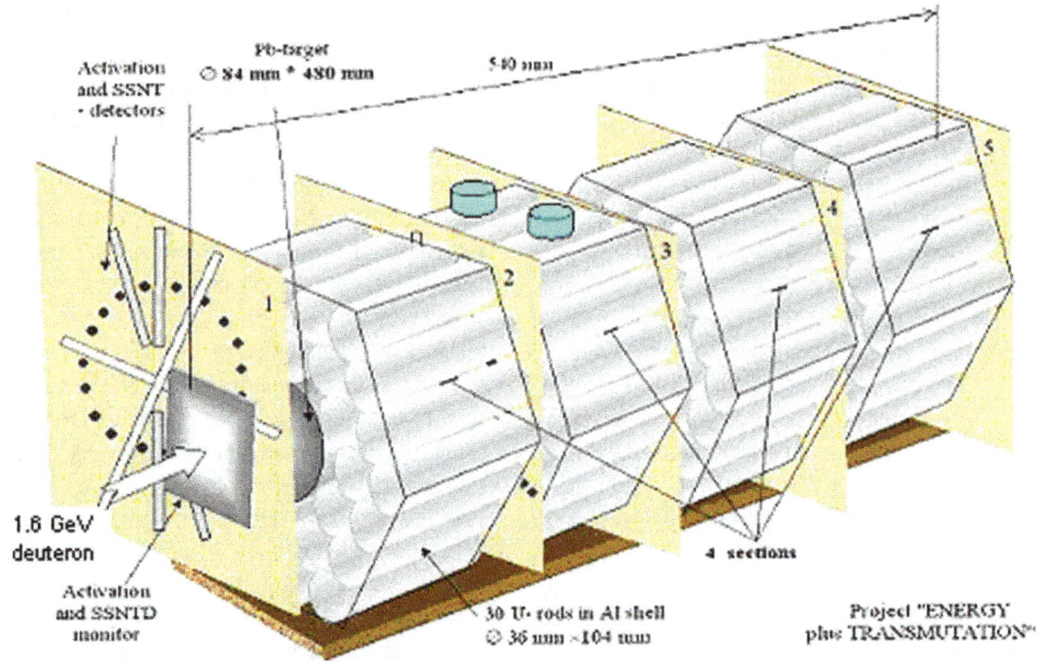

Fig. 6.19 Energy + transmutation setup is shown with massive lead spallation target in the middle of the four hexagonal sections of depleted ^{nat}U assemblies. Activation samples are placed on the top of the sections, and in the middle of any two sections neutron flux monitor detectors are placed. Two radioactive samples like ^{129}I and ^{237}Np are put on the top of the second uranium section [62, 64]

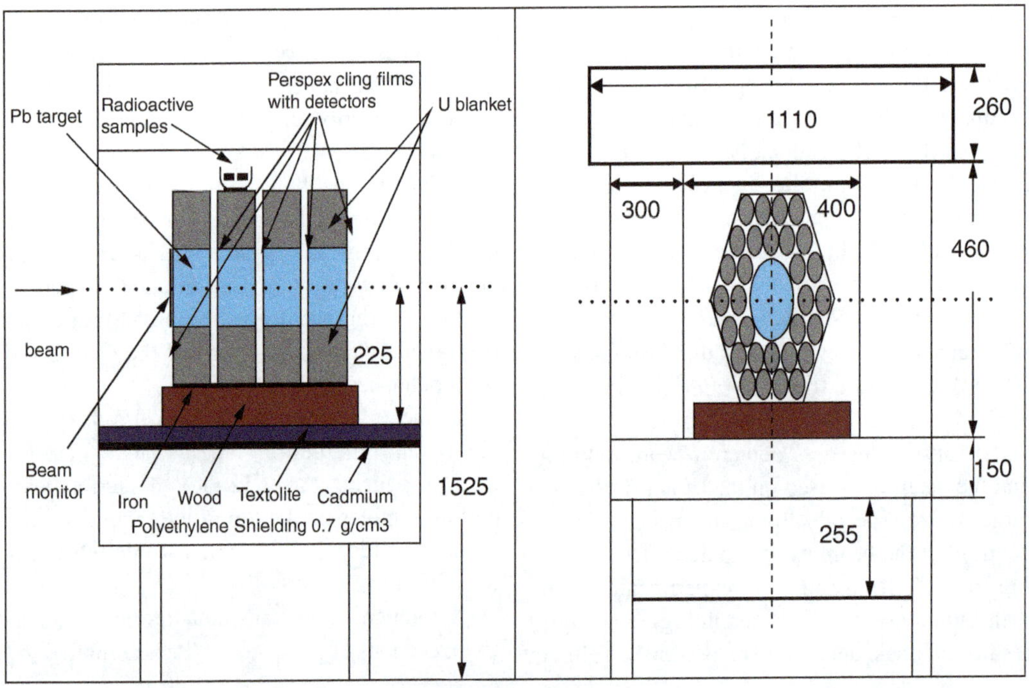

Fig. 6.20 (Left) Side view of energy + transmutation assembly with radiation shielding by polyethylene cover of dimensions $\Delta X \times \Delta Y \times \Delta Z = 100 \times 106 \times 111$ cm^3, 0.1-cm-thick cadmium layer, textolite layer having thickness 3-cm, 9-cm-thick wood support, steel plate of 1 cm thickness, 0.1-cm-thick steel blanket cover, hexagonal uranium blanket around the central lead target. Thin beam monitor is shown in front of the lead target (right) front view of the setup with dimensions used in the simulations by the CASCADE code [62, 64]

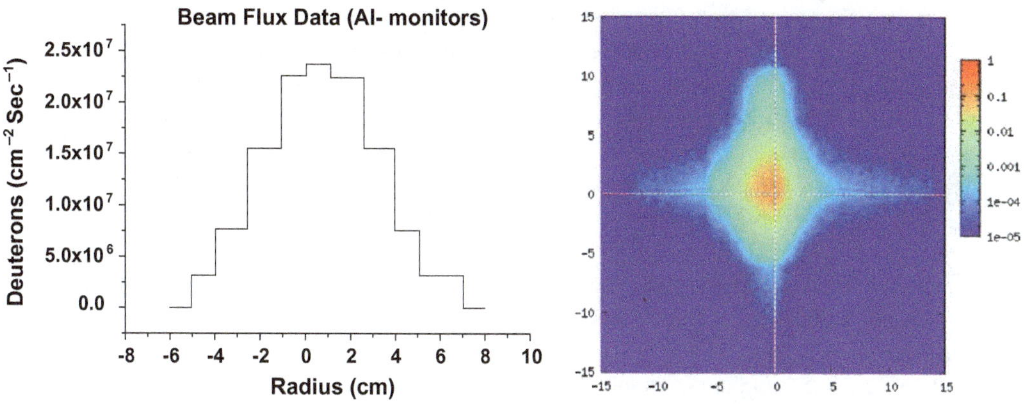

Fig. 6.21 (Left) Beam profile estimated (n/cm²/s) using the Al foil monitors and (right) real beam profile detected using the track detectors [60]. Coordinates of centroid are $x_c = -0.64(3)$ cm and $y_c = 0.39(8)$ cm with respect to the axis of the setup

Fig. 6.22 1.6 GeV deuteron beam profile used in the experiment [60]

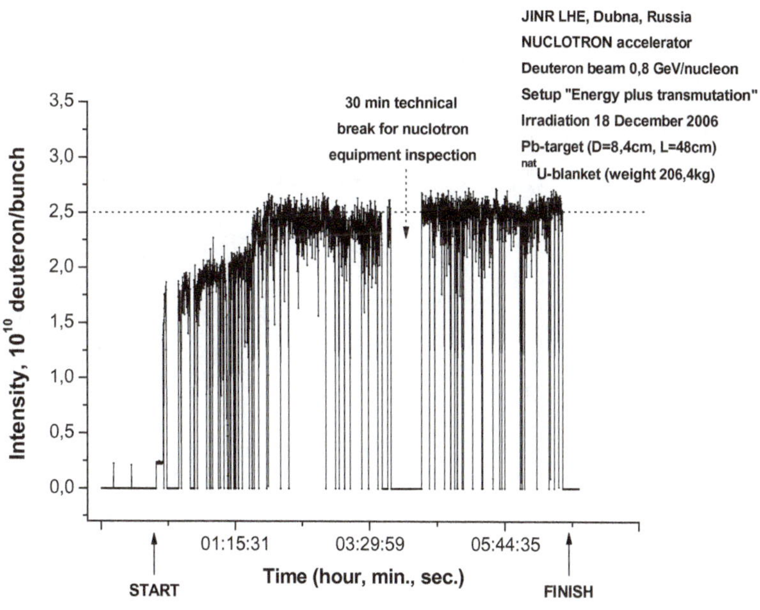

for the $E + T$ setup using data of reaction rates given by Adam et al. [65] for the proton beams of energy ranging from 0.7 to 2 GeV. A detailed discussion of transmutation of RA samples using proton or deuteron beams of different energies has already been presented in Fig. 5.2 where it is inferred that transmutation in an experimental setup of ADSS by energy more than 1 GeV/n does not show much improvement compared to 1 GeV.

6.13 QUINTA—An Experimental Setup

During the Baldin seminars [68], Furman et al. [69] proposed to extend the $E + T$-type setup by adding another uranium assembly and to create an entrance beam window. The new target assembly (TA) along with a mass of 500 kg of natural uranium metal ($d \times l = 30 \times 65$ cm^2) was identified as QUINTA which was irradiated by deuteron beams with energies ranging from 1 to 8 GeV. In the experiments, number of deuterons on the target ranged from 3 to 5 × 10^{13} for each energy. In Fig. 6.24, a schematic view of the setup as been projected. The setup is highly equipped with detector systems for the measurement of beam intensity and the activation samples placed at different positions inside the setup as well as on the top of the uranium assembly. In Fig. 6.25, uranium assembly shown at the center of earlier Fig. 6.24 is shown separately along with entrance beam window placed before each uranium assembly.

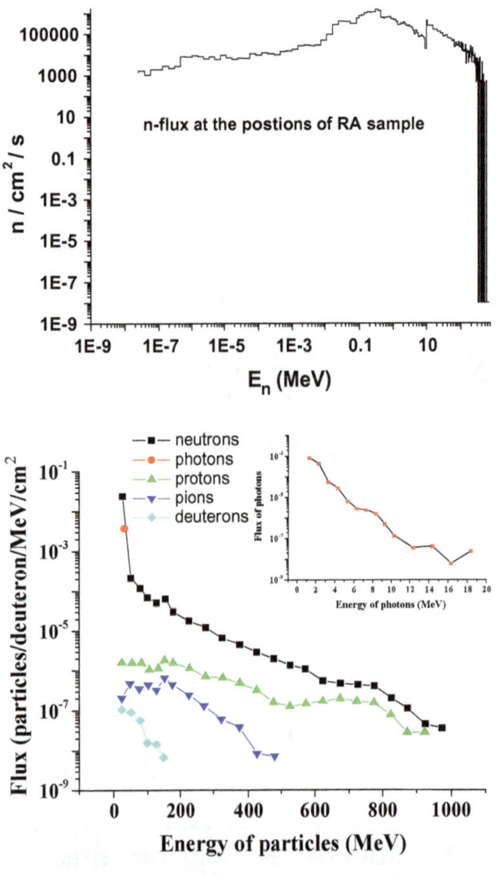

Fig. 6.23 (Up) Neutron flux, as simulated from the CASCADE code, passing through the ^{129}I and ^{237}Np radioactive (RA) samples placed on the top of the second section of U/Pb assembly. Nearly same n-distribution is obtained from the code for the U and Th samples [53]. (Below) Simulation of various produced particles hitting the target, carried out using MCNPX by using the real coordinates of beam [60]. Units of flux and energy scale in case of MCNPX simulations are different to that given in left-side figure

Asquith et al. [70] have measured reaction rates of (n, γ) and (n, xn) reactions produced in activation samples ^{209}Bi and ^{197}Au placed inside the setup irradiated by the deuteron beams of 1 and 4 GeV energy in two experiments separately. Mean values of total beam particles of 1 and 4 GeV energy are estimated to be (1.50 ± 0.16) 10^{13} and (1.94 ± 0.20) 10^{13} respectively. The activation samples were prepared to be sandwiched in between two 100-μm-thick muscovite mica sheets as the SSNTDs for the detection of fission fragments escaping the activation samples. Experimental results of ^{209}Bi (n, xn) reactions with x = 4–7 and ^{197}Au (n, γ) as well as ^{197}Au (x, yn) reactions with y = 2–7 reactions are found to be in good agreement with the calculated results following the MCNPX 2.7 manual [71]. Adam et al. [72] have also presented results of neutron flux measurements in an experiment with QUINTA setup using 4 and 8 GeV deuteron beams and threshold detector reactions, ^{27}Al (n, y_1) ^{24}Na, ^{27}Al (n, y_2) ^{22}Na, and ^{27}Al (n, y_3) ^{7}Be having effective threshold energies as 5, 27 and 119 MeV. Experimental results are found in good agreement with the calculated results at lower neutron energies, i.e., 5–27 and 27–119 MeV in case of 4 GeV deuteron beam.

6.14 MYRRHA and EFIT—A Road Map of ADSS

MYRRHA is abbreviated for **M**ulti-purpose h**Y**brid **R**esearch **R**eactor for **H**igh-tech **A**pplications. In fact, SCK.CEN (Mol, Belgium) had initiated the MYRRHA project in 1998 to pave a road toward construction of an e**X**perimental facility for demonstration of the technical feasibility of **T**ransmutation in **A**ccelerator-**D**riven **S**ystem as the XT-ADS concept [73]. Thus, XT-ADS will be an advance version of MYRRHA. It has assumed inherent security feature by way of subcriticality of the system and to provide extra neutron flux by way of spallation process. The spallation neutrons not only initiate production of prompt and delayed

Table 6.9 Experimental spectrum average reaction rates measured in case of $E + T$ setup using 1.6 GeV deuteron beam for the ^{237}Np, ^{238}U, and ^{232}Th samples. In the parenthesis { }, values of reaction rates for 2.52 GeV deuteron beam are shown for comparison [53]

Reaction and product	^{237}Np	^{238}U	^{232}Th
(n, γ) ^{238}**Np**	1.82(6)E−25 {1.62(8)E−25}	2.92(21)E−26	2.91(14)E−26
$(n, 2n)$ ^{231}**Th**	–	–	1.98(14)E−27
(n, f) ^{143}**Ce**	–	2.8(6)E−28	–
(n, f) ^{132}**Te**	1.61(12)E−27 {2.2(3)E−27}	5.20(26)E−28	–
(n, f) ^{133}**I**	2.40(22)E−27 {2.6(8)E−27}	1.78(22)E−28	–
(n, f) ^{135}**I**	2.21(22)E−27	–	–
(n, f) ^{135}**Xe**	–	1.37(32)E−27	–
(n, f) ^{140}**Ba**	1.54(17)E−27	6.39(18)E−28	–
(n, f) ^{99}**Mo**	–	5.06(86)E−28	6.9(13)E−29
(n, f) ^{97}**Zr**	1.98(12)E−27 {1.88(29)E−27}	–	–

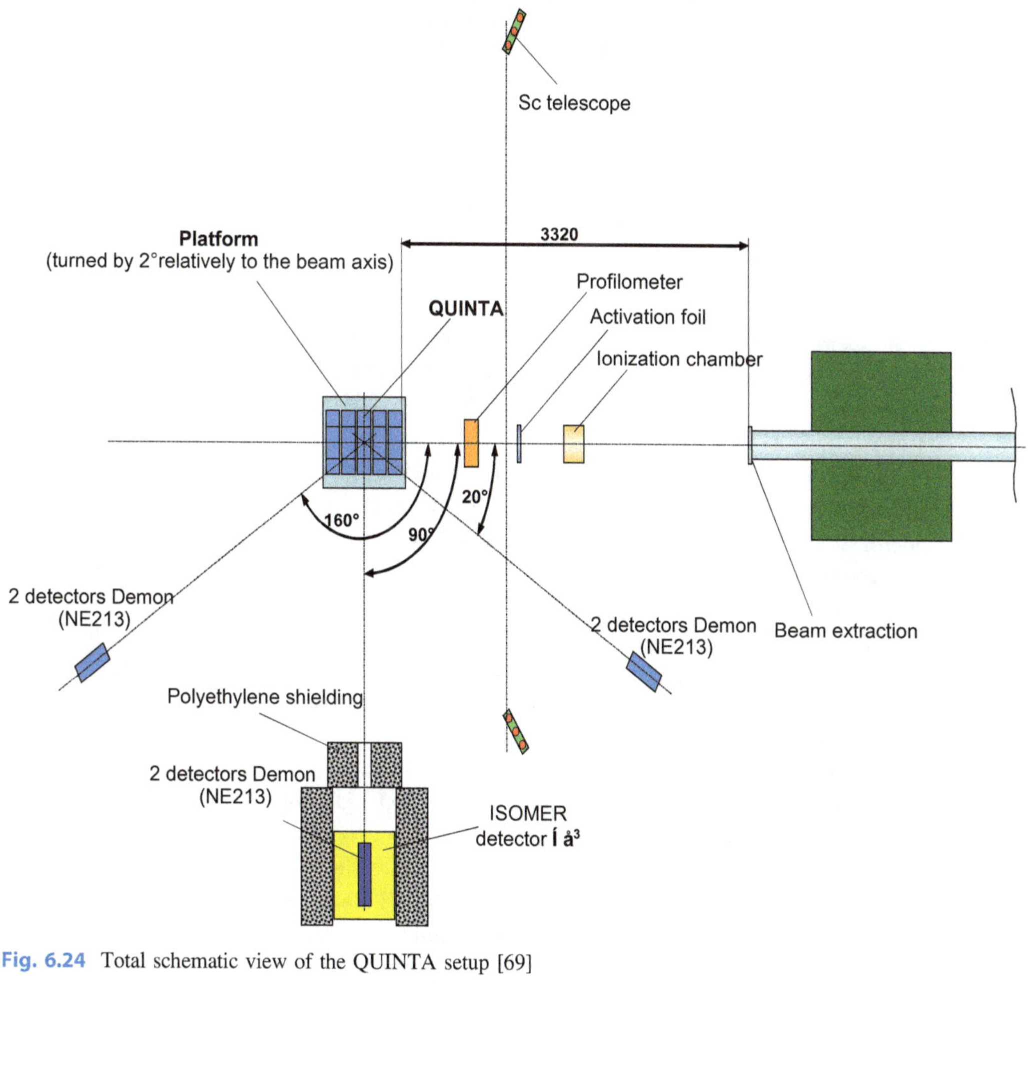

Fig. 6.24 Total schematic view of the QUINTA setup [69]

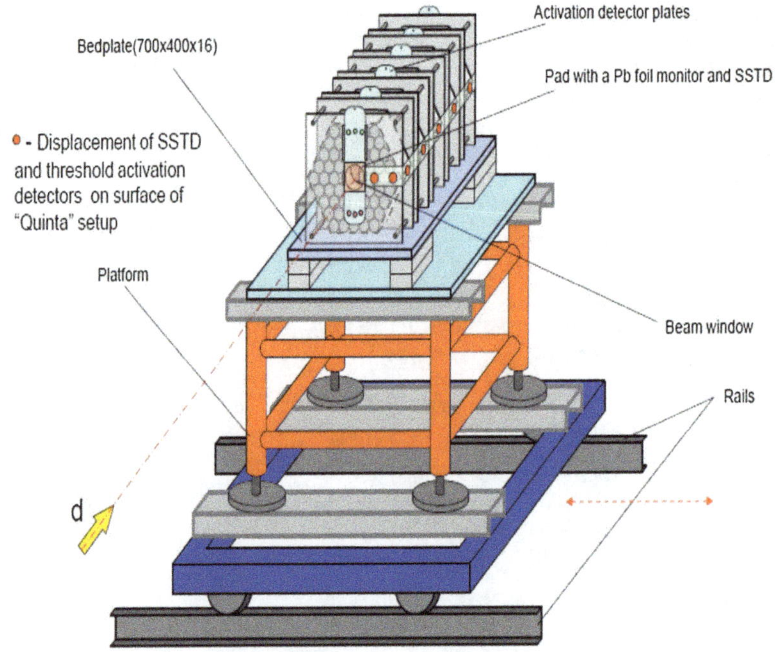

Fig. 6.25 A view of ^{nat}U assembly of five sections with beam window before each section [69]

neutrons but multiply neutrons directly from the material of the reactor. Industrial design of extension of the demonstration design MYRRHA/XT-ADS is called as EFIT. According to Mueller [74], the main objective is to work towards a European Transmutation Demonstration in a step-wise manner, i.e., to provide (i) an advanced design of all the components of an XT-ADS system (ii) a generic conceptual design of a modular 'European Facility for Industrial Transmutation,' EFIT for the long-term objective of the program. Once the demonstration of EFIT starts working, it will need to be characterized with respect to (a) its transmutation efficiency, (b) easy in operation, and (c) its availability level. Comparative design parameters of MYRRHA, XT-ADS, and EFIT are shown in Table 6.10. The entire project is planned to be operational by the year 2020 with the help of several experimental test facilities like liquid heavy metal, Pb–Bi eutectic as target-cum-coolant and GUINEVERE described earlier. In Fig. 6.26, MYRRHA setup is shown with a table of its component assemblies.

6.15 Subcritical Assembly at Dubna (SAD)

Planning of SAD was started with a view of utility of 660 MeV Phasotron accelerator and IBR-30 plutonium reactor at JINR, Dubna. Later, Polanski [75] suggested that MOX fuel ($PuO_2 + UO_2$) is a better fuel than metallic plutonium. In a series of simulations, various fuel combinations were tried [76, 77] for an electronuclear system and their results gave birth to the SAD proposal. Some of the basic parameters of the proposed facility are given in Table 6.11 [78]. In Fig. 6.27, proposed design of the reactor core is shown [79]. But the facility has not been realized up to level of its desired applications.

6.16 BURAN Setup

After a series of experiments such as GAMMA-2 and GAMMA-3, energy plus transmutation ($E + T$), and QUINTA as

Table 6.10 Comparison of design parameters of MYRRHA, XT-ADS, and EFIT setup

Parameters	MYRRHA (2005)	XT-ADS (2009)	EFIT
Goal	**Concept demonstration**	**Prototype transmuter**	**Maximize the transmutation efficiency, easiness of operation and maintenance, high level of availability**
Accelerator, E_p, I	LINAC, 350 MeV, 5 mA	LINAC, 600 MeV, 2.5 mA or as of MYRRHA	LINAC, 800 MeV, 20 mA
Beam entry into the reactor, interface	Top side, windowless	Top side, windowless	As of XT-ADS
Target and coolant	Pb–Bi	Pb–Bi	Pb
Core power	~50 MWth	~57 MWth	Several 100 MWth power
Fuel	MOX (except for a few MA fuel samples)	MOX (except for a few MA fuel assemblies)	Minor actinide fuel
Fuel power density	~1000 W/cm³	700 W/cm³	–
Criticality, k_{eff}	~0.95	~0.95	~0.97

Fig. 6.26 MYRRHA design with a table of components assemblies of the setup [73]

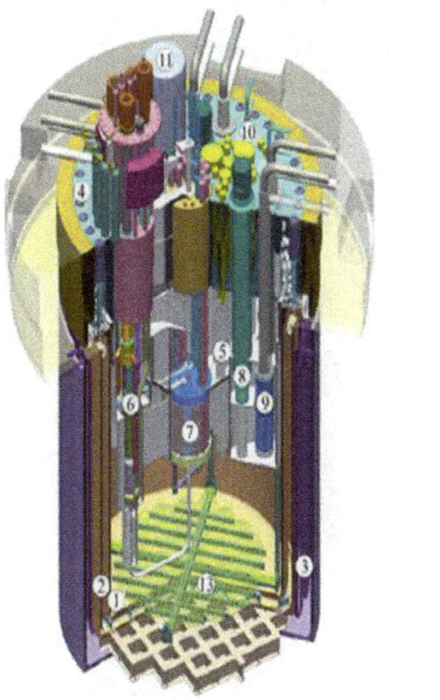

1. inner vessel
2. guard vessel
3. cooling tubes
4. cover
5. diaphragm
6. spallation loop
7. sub-critical core
8. primary pumps
9. primary heat exchangers
10. emergency heat exchangers
11. in-vessel fuel transfer machine
12. in-vessel fuel storage
13. coolant conditioning system

discussed earlier in the chapter, another series of possible experiments with a quasi-infinitely large depleted uranium assembly is being readied for neutron benchmark studies in the name of BURAN at JINR, Dubna. BURAN stands for **big uran**ium assembly. The schematic design of the BURAN setup is shown in Fig. 6.28 comprising of depleted 0.3% ^{235}U + 99.7% ^{238}U

Table 6.11 Proposed design parameters of SAD assembly [78]

Basic component	Basic parameters	Tentative value
Beam	Proton energy	660 MeV
	Beam power	1–2 kW depending on target assemblies
Spallation target	Concentric cylinders of W or Pb (center), solid Pb around followed by beryllium	Be will reduce neutron flux on the blanket
	n/*p*	12.95
	Heat generation	840 W
	Neutron intensity on blanket	1.143×10^{14} n/s
Fuel blanket	MOX	PuO_2 (0.297) + UO_2 (0.703)
	Shape and size	Pallets, 5.95 mm diameter
	No. of fuel units	132–141
	Power	30–114 kW depending on beam and target assembly

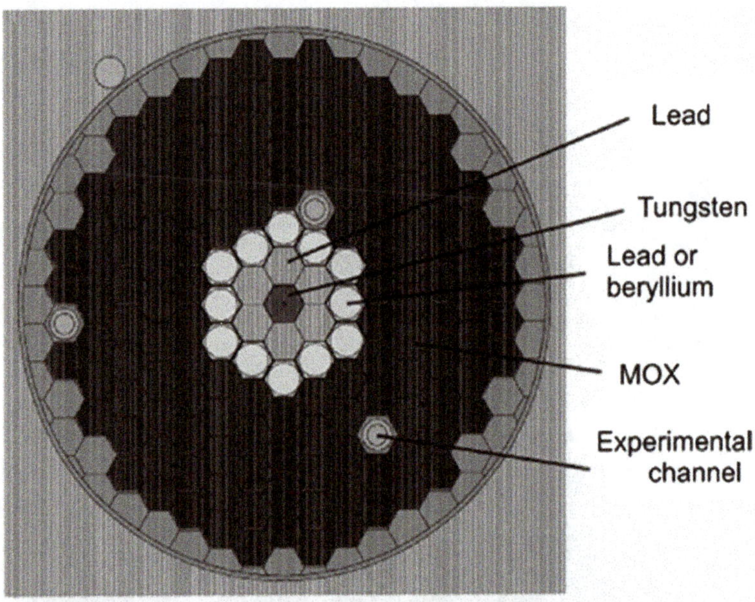

Target	Pb	W+Pb+Be	W+Pb
Fuel Elements	133-141	133	132
k_{eff}	0.952 -0.972	0.974	0.974
E-gain	30 - 50	51	57
Beam power	1 kW	2 kW	2 kW
System power	30-50 kW	102 kW	114 kW

Fig. 6.27 Design of reactor core of SAD [79]

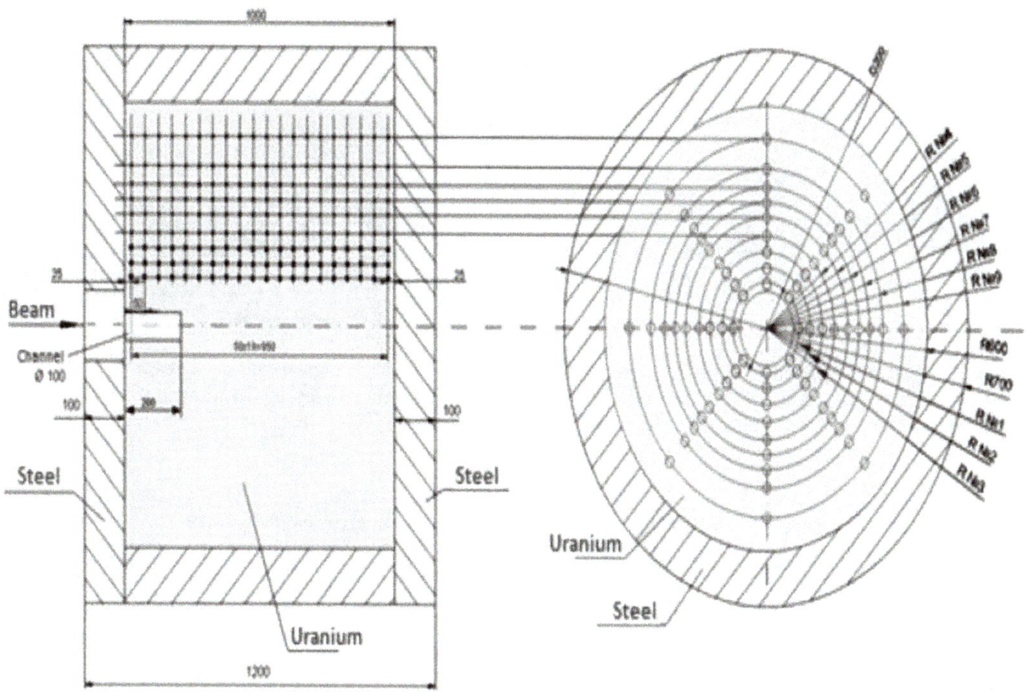

Fig. 6.28 Side view of cylindrical geometry of BURAN experimental setup at JINR, Dubna [80]

assembly of size $d \times l = 120 \times 100$ cm^2 enclosed in 10-cm-thick steel covering. In the center of the cross section of the uranium block, there is a beam opening of 20 cm diameter and its depth is kept moveable at 0, 10, and 20 cm as per necessity.

Neutron spectra are simulated using MCNPX 2.7.0 in different positions of the setup for proton and deuteron beams and shown in Fig. 6.29 (left) and (right), respectively. Calculations show a marginally more peak value of neutron flux in case of deuteron than proton beam of same kinetic energy. Detailed results of MCNPX calculations for proton and deuteron are summarized in Table 6.12 [81], and they are helpful in design of experiments of benchmark related to fissions (n, f) and transmutations by (n, γ) reactions. In the last row of Table 6.12, calculated values of 'beam power gain, BPG' have been shown to be nearly same at all energies in case of the two projectiles [81].

6.17 GEM * STAR Setup

Green **E**nergy **M**ultiplier * **S**ubcritical **T**echnology for **A**lternative **R**eactor is a consortium aimed for (i) addressing world's energy issues, (ii) burn nuclear waste, natural uranium, depleted uranium, thorium and excess weapon-grade plutonium, (iii) to use a superconducting accelerator and Molten Salt fuel to achieve greatly improved safety issues and to address the issues of nuclear waste which are both economically and politically feasible, and (iv) GEM * STAR will even be cheaper than natural gas [82]. As a special feature of having Molten Salt fuel makes GEM * STAR very special as it reduces mechanical fatigues of solid reactor rods which normally occur because of frequent accelerator trips. Molten Salt may be a combination of variety of fuels, and its biggest advantage lies in

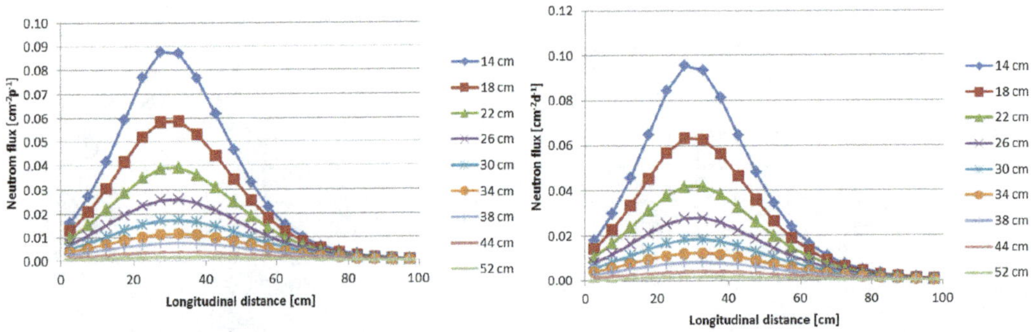

Fig. 6.29 (Left) Neutron flux per incident particle (cm^{-2}) plotted as function of axial distance for 1 GeV proton beam and (right) deuteron beam calculated using the MCNPX [81]. Different curves correspond to the radial distances shown as legends along with the figures

Table 6.12 Average results of calculation of MCNPX for different beam energies [81]

	Protons			Deuterons		
$E_{p(d)}$ (GeV)	1	6	12	1	6	12
Total neutron multiplicity	126	770	1450	125	794	1455
Number $N(n,y)$	70	440	826	70	452	837
Number $N(n,f)$	16	100	183	15	100	183
$K_{BPG} = E_{tot}/E_{p(d)}$	3.82	3.75	3.5	3.82	3.85	3.55

safety against the corrosion issues. It also contains the fission products which can be separated in a cycle. Dynamically, it is highly stable.

6.18 Some More ADS Programs of Asian Countries

China has an ADS program named as C-ADS started from the year 2011 and to be completed in the year 2030. As per its first phase planning, ADS will run with accelerator of 15 MW power LINAC of 1.5 GeV beam with 10 mA current in continuous wave (CW) form. Accelerator will have superconducting cavities for avoiding beam noises which play important role in design of ADSS. The cavities are being developed at Institute of High Energy Physics, Beijing [83]. The project is advancing in convincing way as its couplers are already manufactured.

Japan has proposed its ADS plan for the reduction of burden of disposal of high-level waste by using ADS technology. An ADSS with the thermal power of 800 MW has been proposed, where 250 kg of minor actinide (MA) can be transmuted annually. A superconducting accelerator is being developed for coupling to Pb–Bi eutectic (LBE) spallation target. The corrosive activity of LBE is overcome by controlling oxygen concentration in LBE which forms an oxide layer on its surface. The J-PARC experimental facility and feasibility assessment are already discussed in Sect. 6.5. As the MAs act as fertile fuel, therefore, plutonium is added for the swing. The Japan ADSS is assumed to work at k_{max} = 0.97. Its basic design characteristics are given in Table 6.13 [84].

South Korea is using atomic power, and it has ambitious plan to have 40% energy consumption to be met by atomic power. More than 23

Table 6.13 Core physical parameters of 800 MW Japanese ADS [84]

Parameter	Value	Parameter	Value
Thermal power	800 MW	Initial MA inventory	2,500 kg
Active core diameter	236.6 cm	Effective (k_{eff})	Initial = 0.97, max. = 0.97, mini. = 0.94
Active core height	100.0 cm	Average power density	191 W/cm^3
Initial Pu (inner/outer)	30.0%/ 48.5%	Proton beam energy	1.5 GeV
Total heavy metal inventory	4,115 kg	Proton beam current	Max. = 17.9 mA, mini. = 8.1 mA

reactors are working on its land and it want to add 11 more up to the end of 2024. South Korea is expected to produce 100,000 tons of nuclear waste by the end of the century and for that purpose it will need a safe disposal vault of 20 km^2 in rock caverns and about 500 m underground. According to South Korea's Atomic Energy Research Institute, KAERI, another point of view, pyro-processing technology could reduce waste by 95% compared to 20–50% from the existing reprocessing technology. Country's atomic energy plans are based on uranium fuel, and the country has hardly decided to start an ADSS plan. However, Korea's POSTECH institute has facility for cross-section measurements and KAERI at Seoul had started developing high-current pulsed proton beam accelerator of 100 MeV more than a decade back. According to Karel Samek, iThEC, presently, S. Korea's ADSS plans are getting to start at the university level and SKKU which is planning to produce a high-energy, high-current 5 MW cyclotron for the thorium-based ADSS reactor. The cyclotron will be developed for 1 GeV proton beam for a large number of studies related to ADS activity, and the ADSS may start in the year 2040.

At the same time, S. Korea has shown willingness to work with the USA for the pyro-processing, and according to a report [85] US has agreed to work on the technique jointly.

In Asia, India has been pioneer in initiating its atomic energy programs of peaceful applications of atomic energy in the year 1954. First experimental reactor Apsara, 1 MWt, was started soon after. It has a long-term objective primarily for utilization of its major resource of thorium which is largely explored on its west and east sea shores. Basically, it has three-stage strategy of atomic energy sustenance. It has developed fast neutron spectrum experimental reactors for developing fast reactors and radiography of materials. It has upgraded its 14 MeV neutron source for higher current ion source also. Its Department of Atomic Energy is engaged in developing Advanced Heavy Water Reactor (AHWR) which aims at developing expertise for thorium utilization and demonstrating advanced safety concepts. Mixed thoria–urania and thoria–plutonia are the candidate fuels for the AHWR. A high-current (30 mA) proton beam LINAC is under development at BARC. Kapoor [86] has given road map of Indian ADS. During Jaipur workshop on Physics of ADSS [87], a spurt in ADS activity in India was noticed from the point of accelerator development and papers presented on developing a dynamics of Pb–Bi eutectic flow loop, developing CASCADE code for simulations and various other collaborative experiments in direction of planning of ADSS. In its stage III of atomic energy, India plans for enhancing breeding of ^{233}U from thorium for utilization in fast reactors. PURNIMA reactor is redesigned to be run by (*d*, *t*) neutron source in subcritical mode, and the efforts will grow to develop understanding of coupling of an accelerator to a reactor and neutron utilization [88]. During this period, a road map of India's 'fast-thermal ADS' is also prepared for initiating the related activities at different places where expertise is being readied.

Explicitly, there is no dedicated ADSS plan presently working in India, but various programs are focused to gather working knowledge from its various academic and engineering projects in phase with several other nuclear countries have been doing for future. It may be inferred that ADS is an ambitious project from the point of availability of financial resources.

References

1. Andriamonje, S., et al.: Experimental determination of the energy generated in nuclear cascades by a high energy beam. Phys. Lett. B **348**, 697 (1995)
2. Rubbia, C.: Resonance enhanced neutron captures for element activation and waste transmutation. CERN Internal Report, CERN/LHC/97–04 (EET), June 1997
3. Abanades, A., et al.: CERN-SL-2001-033-EET Report (2001)
4. Ferrari, A., Sala, P.R.: The physics of high energy reactions' lectures at ICTP. In: Trieste Workshop, Apr 15–May 1996. Also, Fasso, A., Ferrai, A., et al.: A comparison of FLUKA simulations with measurements of fluence and dose in calorimeter structures. Nucl. Instrum. Meth. **A332**, 459 (1993)
5. European Commission, Community Research, Project Report, Nuclear Science and Technology, Neutron Driven Nuclear Transmutation by Adiabatic Resonance Crossing (TARC), EUR 19117 EN. European Commission, Luxemberg (1999)
6. OECD/NEA Data bank, Issy-les Moulineaux, France (1994)
7. Wang, T., et al.: Measurements of the total cross section and resonance parameters of molybdenum using pulsed neutrons generated by the electron linac. Nucl. Instrm. Method. Phys. Res. **B266**, 561 (2008)
8. Zugec, P., et al.: Experimental neutron capture data of 58Ni from the CERN n_TOF facility. Phys. Rev. **C89**, 014605 (2014)
9. Guerrero, C., et al.: Performance of the neutron time-of-flight facility n_TOF at CERN. Eur. Phys. J. **49A**, 27 (2013) and Tobias J.W., et al.: High precision measurements of the ^{238}U (n, γ) cross sections at the n_ToF facility CERN. In: Proceedings of NEMEA-7/CIELO Workshop, Copy right OECD, pp. 149–157 (2014)
10. Larson, N.M.: RSICC Peripheral Shielding Routine Collection SAMMY-M2a: A Code System for Multilevel R-Matrix Fits to Neutron Data Using Bayes' Equations (PSR-158, SAMMY-M2a). Oak Ridge National Laboratory (1999)
11. Lane, A.M., Thomas, R.G.: R-matrix theory of nuclear reactions. Rev. Mod. Phys. **30**, 257 (1958)
12. Mughabghab, S.F.: Atlas of Neutron Resonances, 5th edn. National Data Center, BNL, Upton, USA
13. Zugec, P., et al.: Data available in article, arXiv: 1402.1032v1 [nucl-ex], 5 Feb 2014
14. Lederer, C., et al.: Neutron Capture Cross Section of Unstable 63Ni: Implications for Stellar Nucleosynthesis (2013). arXiv:1304.3310v1 [nucl-ex], 11 Apr 2013
15. Abbondanno, U., et al.: Neutron capture cross section measurement of 151Sm at the CERN neutron time of flight facility (n_TOF). Phys. Rev. Lett. **93**, 161103 (2004)
16. Fujii, K., et al.: International Conference on Nuclear Data for Science and Technology (2007). https://doi.org/10.1051/ndata:07519, http://nd2007.edpsciences.org or http://dx.doi.org/10.1051/ndata:07519
17. Domingo-Pardo, C., et al.: New measurement of neutron capture resonances of 209Bi. Phys. Rev. C **74**, 025807 (2006)
18. Aert, G., et al.: Neutron capture cross section of 232Th measured at the n_TOF facility at CERN in the unresolved resonance region up to 1 MeV, Phys. Rev. C **73**, 054610 (2006), also, http://hal.in2p3.fr/in2p3-00082118
19. Tobias, J.W., et al.: High-precision measurement of the ^{238}U(n,g) cross-section at the n_TOF facility, CERN, NEMEA-7/CIELO Workshop Proceedings, p. 149 (2014)
20. Derrien, H., et al.: Neutron Resonance Parameters of ^{238}U and the Calculated Cross sections from the Reich Moore Analysis of Experimental Data in the Neutron Energy Range from 0 keV to 20 keV, ORNL/TM-2005/241 (2005)
21. Paradela, C., et al.: Proceedings of the Final Scientific EFNUDAT Workshop. In: Chiaveri, E. (ed.) p. 33, 30 Aug–2 Sept 2010. CERN, Geneva (2010). Also, Phys. Rev. **C82**, 034601 (2010)
22. González, E.: On Behalf of the *n*_ToF Collaboration: Neutron Cross-Section for P&T and ADS at the n_ToF Facility at CERN (2010). https://indico.cern.ch/event/83067/attachments/1070068/1525949/Proceedings_EFNUDAT_August2010_REV_FINAL.pdf
23. IREN Project. Intense Resonance Neutron Source. Compiled by A.K. Krasnykh, V.L. Lomidze, A.V. Novokhatsky, Yu, P. Popov, W.I. Furman. Frank Laboratory of Neutron Physics, JINR, Dubna, 1994 and see V.L. Aksenov et al. Acta Phys. Hung. **75** (i-4), 341 (1994)
24. Antropov, V., Dolya, S., Kaminsky, A., Krasnykh, A., Laziev, E., Pjataev, V., Shvets, V., Sumbaev, A.: IREN Test Facility at JINR, LINAC 96, pp. 505–507. https://accelconf.web.cern.ch/accelconf/l96/PAPERS/TUP68.PDF
25. Meshkov, I., Sisakian, A., Trubnikov, G.: IPAC 10, The First International Particle Accelerator Conference, Kyoto, Japan, 23–28 May 2010
26. Logatchev, P.V., et al.: Intense Resonance Neutron Source (IREN) A New Pulsed Source for Nuclear Physical and Applied Applications, https://www.researchgate.net/publication/242182317
27. Jeneva, N.B., et al.: Future Neutron Data Activity on the Neutron Source IREN (2011). http://www.iaea.

org/inis/collection/NCLCollectionStore/_Public/48/009/48009766.pdf
28. Advance Materials for Energy. http://www.ga.com/advanced-materials
29. Persic, A.: TRIGLAV—A Program Package for Research Reactor Calculations. J. Stefan Report, US-DP-7862 (1998)
30. Persic, A., Ravnik, M., Zagar, T.: European Nuclear Society, Berne (Switzerland), Belgian Nuclear Society (Belgium), International Atomic Energy Agency, Vienna (Austria), 217 p, pp. 187–191 (1999). https://inis.iaea.org/search/search.aspx?orig_q=RN:32034894
31. Karch, J., et al.: Performance of the solid deuterium ultra-cold neutron source at the pulsed reactor TRIGA Mainz. Eur. Phys. J. **A50**, 78 (2014)
32. Nagamiya, S.: JAER-KEK joint project on high intensity proton accelerators. In: Proceedings of International Conference on Radiation Shielding (ICRS-9), Tsukuba, 18–21 Oct 1999. J. Nucl. Sci. Technol. (Supplement 1) 40 (2000)
33. JAERI-Tech 2000-003/KEK Report 99-5 (2000) (In Japanese). See also The Joint Project for High-Intensity Proton Accelerators, JAERI-Tech 056/KEK Report 99-4 (1999). Also see, https://eventbooking.stfc.ac.uk/uploads/ioppab14/plostinar.pdf
34. Teshigawara, M., Watanabe, N., Kai, T., Nakashima, H., Nagao, T., Oyama, Y., Ikeda, Y., Kosako, K.: Neutronic Study on the JAERI 5 MW Spalltion Neutron Source-Neutronic Performance of the Reference Target-Moderator-Reflector System and the Target Shape/Size Effects, JAERI-Research 99-020, Mar 1999
35. Maekawa, F., et al.: First neutron production utilizing J-PARC pulsed spallation neutron source JSNS and neutronic performance demonstrated. Nucl. Instrm. Meth. Phys. Res. **620**, 159 (2010)
36. Sakasai K., et al.: Materials and Life Science Experimental Facility at the Japan Proton Accelerator Research Complex III: Neutron Devices and Computational and Sample Environments. Quantum Beam Science 1, 10 (2017) and see, UCN at KEK: Ultra Cold Neutron Source. https://www.kek.jp/en/Facility/IPNS/UltraColdNeutronFacility/
37. Masatoshi F., et al.: 1-MW pulsed spallation neutron source (JSNS) at J-PARC. J. Neutron News **22**, 15 (2011) and https://www.kek.jp/en/Facility/
38. Paranov, D.V., Paranova, E.D.: Generation IV Concepts USSR and Russia. http://dx.doi.org/10.1016/B978-0-08-100149-300012-4
39. Soule, R., et al.: Neutronic studies in support of accelerator-driven systems: The MUSE experiments in the MASURCA facility. Nucl. Sci. Eng. **148**, 124 (2004)
40. Berglöf, C., et al.: Pulsed Neutron Source Reference Measurements in the Subcritical Experiment YALINA-booster. ADS/ET-5. file:///G:/Book%20on%2014072017/Book%20on%20ADS27032017/Chapter%207KK/YELINA%20booster.pdf
41. Persson, C.-M.: Reactivity Determination and Monte Carlo Simulation of the Subcritical Reactor Experiment—"Yalina". KTH Engineering Science. ISBN 91-7178-015-7, Universitetsservice US-AB, Stockholm (2005)
42. Sjöstrand, N.G.: Measurement on a subcritical reactor using a pulsed neutron source, Arkiv för fysik **11**, 13 (1956)
43. Billebaud, A., Baeten, P., Aït Abderrahim, H., Ban, G., Baylac, M., et al.: The GUINEVERE project for accelerator driven system physics. In: International Conference GLOBAL 2009. The Nuclear Fuel Cycle: Sustainable Options Industrial Perspectives, Sep 2009, Paris, France, pp. 1809–1815 (2009). Also, HAL Id: in2p3-00414431 http://hal.in2p3.fr/in2p3-00414431
44. Baylac, M.: Some high lights of ADS programs in Europe. In: 1st International Workshop on Accelerator-Driven Sub-Critical Systems & Thorium Utilization, 27–29 Sept 2010, Virginia Tech. www1.phys.vt.edu/~kimballton/gem-star/workshop/presentations/baylac.pdf
45. Iyengar, P.K., Srinivasan, M., Nargundkar, V.R., Chandramoleshwar, K., Subha Rao, K., Pasupathy, C.S., Das, S., Basu, T.K., Job, P.K.: PURNIMA fast critical facility, experiments and results. In: Proceedings of the Indo–Soviet Seminar on Fast Reactors, Kalpakkam, India, pp. 249–268 (1972)
46. Basu, T.K., Radha, E., Reddy, C.P., Ganesan, S.: (evaluators) NEA/NSC/DOC/(95)03/I, vol. 1, PU-COMP-FAST-004, 30 Sept 2012
47. Srinivasan, M., Chandramoleshwar, K., Job, P.K., Pasupathy, C.S., Subba Rao, K., Ray, A.K., Patil, R.K., Ramanujam, A.: Safety Analysis Report of Purnima-II: BeO Reflected U-233 Uranyl Nitrate Solution Reactor Experiment. BARC/I-790, Bhabha Atomic Research Centre (1984)
48. Srinivasan, M., Chandramoleshwar, K., et al.: ^{233}U fueled homogeneous reactor PURNIMA-II: design features and experimental results. Paper Presented at the Seminar on Fast Reactor Fuel Cycle, IGCAR, Kalpakkam, 10–12 Feb 1986
49. Basu, T.K., Radha, E., Reddy, C.P., Rasheed, K.K., Ganesan, S. (evaluators): PURNIMA-II: U-233 Uranyl Nitrate Solution Reactor with Beryllium Oxide Reflector, NEA/NSC/DOC/(95)03/V, vol. V, U233-SOL-THERM-007
50. Kumar, R., Degweker, S.B., Patel, T., Bisnoi, S., Adhikari, P.S.: Reactor physics experiments in PURNIMA sub critical facility coupled with 14 MeV neutron source. In: Proceedings of the Second International Workshop on Accelerator-Driven Sub-critical Systems and Thorium Utilization (2011). Also see https://inis.iaea.org/search/searchsinglerecord.aspx?recordsFor=SingleRecord&RN=44027640
51. Radha, E., Reddy, C.P., Ganesan, S., Srinivasan, G., Ramalingam, P.V., Raj, B. (evaluators): Kalpakam Mini (KAMINI) Reactor: Beryllium-Oxide Reflected

Water-Moderated ^{233}U-Fueled Reactor. NEA/NSC/DOC/(95)03/V, vol. V, U233-MET-THERM-001
52. Sinha, A., et al.: BRAHMMA: A compact experimental accelerator driven subcritical facility using D-T/D-D neutron source. Ann. Nucl. Energy **75**, 590 (2015)
53. Sharma, M.: Transmutation of Long-Lived Isotopes of Conventional Reactors Using A.D.S. Concept. Ph. D. thesis, University of Rajasthan, Jaipur (2010)
54. Barashenkov, V.S.: Monte Carlo Simulation of ionization and Nuclear Processes Initiated by Hadron an Ion Beams in Media. Comput. Phys. Commun. **126**, 28 (2000)
55. Adam, J., et al.: Measurements of the neutron fluence on the spallation source at dubna. KERNTECHNIK (J. Nucl. Eng. Energ. Syst. Radiat. Radiol. Prot.) **70** 127 (2005)
56. Adam, J., Bhatia, C., Katovsky, K., Kumar, V., Majerle, M., Pronskikh, V.S., Khilmanovich, A.M., Martsynkevich, B.A., Zhuk, I.V., Golovatiouk, V.M., Westmeier, W., Solnyshkin, A.A., Tsoupko-Sitnikov, V.M., Potapenko, A.S.: A study of reaction rates of (n,f), (n,γ) and (n,2n) reactions in ^{nat}U and ^{232}Th by the neutron fluence produced in the graphite set-up (GAMMA-3) irradiated by 2.33 GeV deuteron beam. Eur. Phys. J. A **47**, 85 (2011)
57. Bhatia, C., Kumar, V.: Book: Role of (*n*, *xn*) Reactions in Accelerator Driven Sub-critical Systems. Lap-Lambert, Germany (2011). ISBN 978-3-8465-0684-4
58. Abanades, A., Aleixandre, J., Andriamonje, S., et al.: Results from the TARC experiment: Spallation neutron phenomenology in lead and neutron-driven nuclear transmutation by adiabatic resonance crossing. Nucl. Inst. Meth. A **478**, 577 (2002)
59. Hendricks, J.S., et al.: MCNPX, Version 2.6.c, LA-UR-06-7991 (2006)
60. Adam, J., et al.: A study of nuclear transmutation of Th and ^{nat}U with neutrons produced in a Pb target and U blanket irradiated by 1.6 GeV deuterons. Eur. Phys. J. **A43**, 159 (2010)
61. Bhatia, C., Adam, J., Kumar, V., Katovsky, K., Majerle, M., Solnyshkin, A.A., Tsoupko-Sitnikov, V. M.: A Study of non-elastic reaction rates for the A.D. S. materials in the environment of spallation neutrons produced by 1.6 GeV d-beam. Appl. Radiat. Isot. **70**, 1254 (2012)
62. Krivopustov, M.I., Chultem, D.: Experiments on electronuclear technology and transmutation of nuclear waste using synchrophasotron beams. JINR News Dubna **3**, 27 (1998)
63. Krivopustov, M.I., Chultem, D., Tumendelger, T., et al.: Modeling of the electronuclear method of energy production and study of radioactive waste transmutation using a proton beam of the JINR synchrophasotron/nuclotron: research program of the laboratory of high energies. JINR. In: Baldin, A.M. (ed.) JINR, 99–266. Dubna, pp. 135–139 (1999). Book of Abstracts of the 3rd International Conference on Accelerator Driven Transmutation Technologies and Applications, (1999) Praha, Czech Republic, 7–11 June 1999, p. 55
64. Barashenkov, V.S., Kumawat, H., Lobanova, V.A., Kumar, V.: Mathematical model of the electronuclear set-up on the beam of JINR Synchrotron. Nucl. Instr. Meth. Phys. Res. B, **217**, 352 (2004)
65. Adam, J., et al.: Transmutation of ^{129}I, ^{237}Np, ^{238}Pu, ^{239}Pu, and ^{241}Am using neutrons produced in target-blanket system 'Energy plus Transmutation' by relativistic protons. Pramana J. Phys. **68**, 201 (2007)
66. MONC.: Monte Carlo Nucleon Transport Code. http://bts.barc.gov.in/AuthStatic/auth_static.php/npd/monc/index.html. Accessed on 19 Dec 2016 (BARC internal access only)
67. Kumar, V., Tundwal, A., Katovsky, K., Sharma, M.: A review of nuclear waste transmutation using accelerator beams up to several GeV. Talk Presented at 17th International Scientific Conference on EPE-2016, Praha, Czech Republic (2016). http://ieeexplore.ieee.org/stamp/stamp.jsp?tp=&arnumber=7521826&isnumber=7520888
68. Baldin, A.A., Belov, E.M., Galanin, M.V., et al.: Nuclear relativistic technologies (RNT) for energy production and utilization of spent nuclear fuel. Results of first experiments on physics justification of RNT. Particles Nucl. Lett. **6**, 1007 (2011) (in Russian). JINR Preprint **E1-2011-24**, Dubna (2011)
69. Furman, W., et al.: Recent results of the study of ADS with 500 kg natural uranium target assembly QUINTA irradiated by deuterons with energies from 1 to 8 GeV at JINR NUCLOTRON. 2012 Proceedings of Science (Baldin ISHEPP XXI 086). http://pos.sissa.it/archive/conferences/173/086/Baldin%20ISHEPP%20XXI_086.pdf
70. Asquith, N.L., et al.: Activation of ^{197}Au and ^{209}Bi in a fast spectrum sub-critical assembly composed of 500 kg natural uranium irradiated with 1 and 4 GeV deuterons. Ann. Nucl. Energy **63**, 742 (2014)
71. Pelowitz, D.B.: MCNPX User's Manual Version 2.7.0, LA-CP-11-00438 (2011)
72. Adam, J., et al.: Measurement of the high energy neutron flux on the surface of the natural uranium target assembly QUINTA irradiated by deuterons of 4 and 8 GeV energy. Phys. Procedia **80**, 94 (2015)
73. De Bruyn, D.: Engineering Design of the MYRRHA (Design Evolution from MYRRHA to XT-ADS) Part IX, School on Physics, Technology and Applications of Accelerator Driven Systems (ADS), 19–30 Nov 2007. ICTP, 1858-14 and Abderrahim, H.A., et al.: MyRRHA–A Multipurpose Fast Spectrum Research Reactor. Energ. Convers. Manag. **63**, 4 (2012)
74. Mueller, A.C.: Transmutation of Nuclear Waste and the Future MYRRHA Demonstrator. https://arxiv.org/pdf/1210.4297

75. Polanski, A.: Monte Carlo modeling of electronuclear processes in experimental accelerator driven systems. Acta Phys. Polonica, **B11**(1), 95 (2000)
76. Barashenkov, V.S., Puzynin, I.V., Polanski, A.: Mathematical experiments with electronuclear systems. J. Comput. Meth. Sci. Eng. **2**(1–2), 5 (2002)
77. Barashenkov, V.S., Polanski, A., et al.: Fast sub-critical assembly with MOX fuel for research on nuclear waste transmutation. Belarus NSA J. Phys. Tech. **3**, 150–153 (2001)
78. Shvetsov, V.N., et al.: Sub critical assembly at Dubna (SAD): coupling all major components of accelerator driven subcritical system (ADS) for nuclear waste incineration. In: International Symposium on Utilization of Accelerators, Dubrovniik, Croatia, 5–9 June 2005. http://www-naweb.iaea.org/napc/physics/meetings/dubrovnik2005/papers/82.pdf
79. Polanski, A.: Simulations of accelerator driven systems (ADS). In: Joint IAEA-ICTP Workshop on Nuclear Reaction Data for Advanced Reactor Technologies. https://www-nds.iaea.org/workshops/smr1944/Participants/28-May.../POLANSKI.ppt
80. Tichý, P., Suchopár, M.: Future usage of quasi-infinite depleted uranium target (BURAN) for benchmark studies. In: Proceedings of Science, XXII International Baldin Seminar on High Energy Physics Problems, JINR, Dubna, Russia, 15–20 Sept 2014. https://pos.sissa.it/225/065/pdf
81. Furman, W.: Results of experiments 2012–2013 with massive uranium target setup QUINTA at nuclotron and plans for 2014–2016. In: ISINN XXI, Alushta, Ukraine, May 20–25 2013
82. Bowman, C., et al.: GEM*STAR* consortium staging and demonstration. In: 1st International Workshop on ADSS and Thorium Utilization, Blacksburg, VA, 27–29 Sept 2010
83. Huang, T.M., et al.: High power input couplers for C-ADS. In: Proceedings of SRF 2015, Whistler, BC, Canada, THPB101, p. 1383 (2015). ISBN 978-3-95450-178-6
84. Tsujimoto, K., et al.: Neutronics design for lead-bismuth cooled accelerator—driven system for transmutation of minor actinide. J. Nucl. Sci. Technol. **41**, 21 (2004)
85. Nuclear Waste a Growing Headache for South Korea. https://www.ndtv.com/world-news/nuclear-waste-a-growing-headache-for-south-korea-517345
86. Kapoor, S.S.: Accelerator-driven sub-critical reactor system (ADS) for nuclear energy generation. Pramana J. Phys. **59**, 941 (2002)
87. Kumar, V. et al. (ed.).: Proceedings of Workshop on Physics of Accelerator Driven Subcritical System. Pramana J. Phys. **68** (2007) (complete issue)
88. Banerjee, S.: Physics of ADS for energy and transmutation. In: Inaugural Talk at the Workshop on Physics of ADS, Jaipur, India, 23–25 Jan 2006

7 Radiation Damage and Development of a MC Software Tool

The chapter comprises innovative efforts taken in case of defining radiation damage. Older models and codes used for the estimation of radiation damage are considered as a matter of discussion elsewhere. In fact, simulation codes and models of calculation of data required for designing new energy systems, have more concern of comparison with the new experimental data which is being collected world over as described in earlier chapters. For computation of radiation damage up to several MeV energy, atomic collision cross section play much vital role than consideration of passage of the products of inelastic collision at these energies. In case of radiation damage by passage of gamma radiation estimation of single vacancies and interstitials is highly desired in design and modeling of several electronic devices. This requires attention to the subject of radiation damage too. Nevertheless, role of dynamical models including kinetic Monte Carlo have more vital role to play in future activities of applications of radiation in the field of medical science.

7.1 Radiation Damage—An Important Issue

In case of a power reactor issue of radiation damage is pertinently related to the structure material, core structure and shielding; whether the material remains intact or not without losing its functionality. Integrity of the fuel rods or fuel cells, coolant pipes, material used to contain the inner or the outer cores is important under the severe irradiation by neutrons, gamma, charged particles, harsh temperature conditions, and the flow rate of the coolant, etc. This becomes a more important issue in case of fusion reactors and the ADSS where the inner wall of a fusion reactor and other structure materials in case of both face many times more intense neutron fluence. Neutrons not only activate the material but also multiply by way of (*n*, *xnyp*) kind of reactions.

Radiation damage so far is measured in terms of displacements per atom (DPA) and according to one estimate [1] in case of thermal reactors the core sees a damage up to 20 DPA and it will rise up to $\sim$120 in a fast reactor and even more in case of ADSS. In case of future generation IV reactors, it will be an order of magnitude higher particularly on the inner wall of a fusion reactor. Physical effects arising due to irradiation shown in Fig. 7.1 indicate that vacancies, dislocations, vacancy loops, voids, clustering, and microstructures, etc., become the cause of weakening of the materials. A material may become amorphous and swell crossing the designed boundaries. Embrittlement and corrosion of the metals are the physically visible effects. Formations of radicals in liquids causing many chemical reactions occur also due to soft radiation passage.

In Fig. 7.2, radiation damage of core structure materials of different nuclear reactors has been summarily shown.

V. K. Verma and K. Katovsky, *Spent Nuclear Fuel and Accelerator-Driven Subcritical Systems*, Green Energy and Technology, https://doi.org/10.1007/978-981-10-7503-2_7

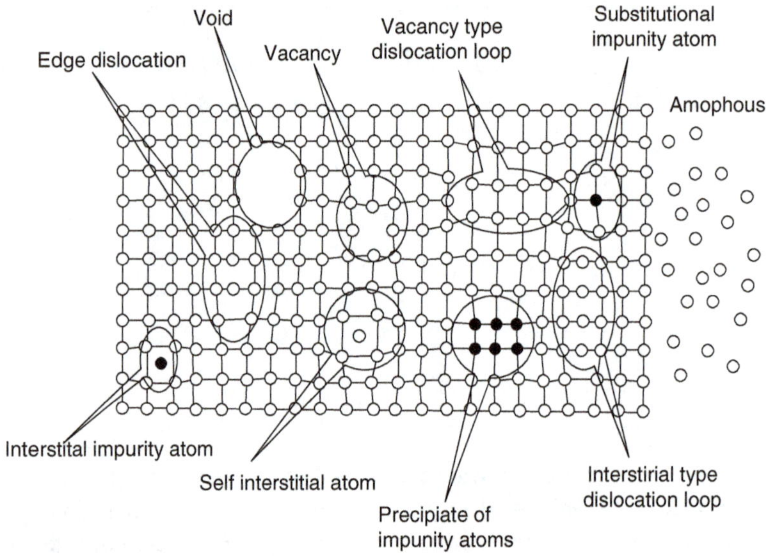

Fig. 7.1 Physical effects of radiation damage in the bulk of matter [1]

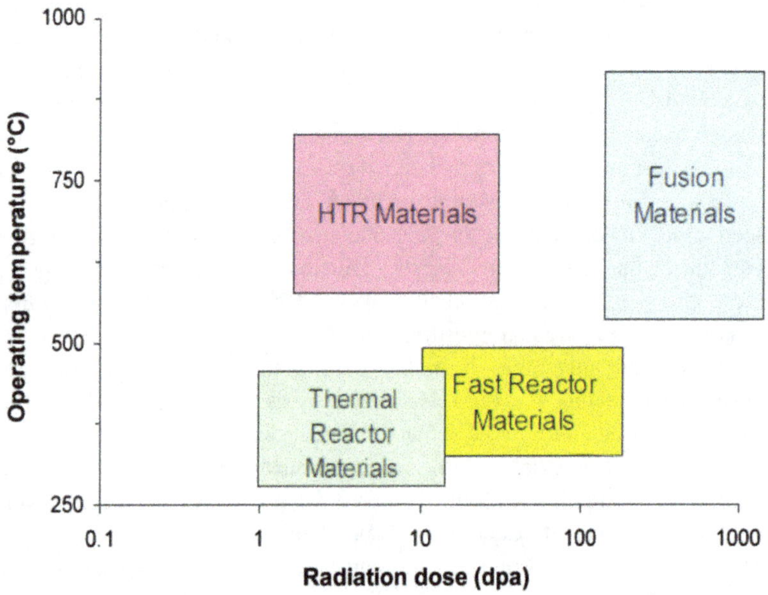

Fig. 7.2 Operating conditions of structure materials of different reactors [1]

Issues related to development of the subject of radiation damage and modern requirements of radiation resistant materials have been deliberated primarily in Ref. [2]. In fact, the subject of development of radiation resistant materials needs utmost attention of the initiatives of the new definitions and fundamental theoretical developments. In the chapter, such initiatives have been discussed which need further intensive research works.

7.1.1 Radiation Effects in Materials

Radiation damage has been an important issue of a power reactor and both its material and the structures are strongly built looking at the effects of radiation. Accelerator beam pipe and the materials in immediate surrounding are the second irradiated materials, and they are usually handled very carefully. Effects like 'Channeling and sputtering' became the subjects of high attraction in the area of material studies in early 70s and 80s of last century. In the last two decades, material modification, synthesis, upgradation of materials have attracted attention heavily and at the same time they have made strong impact on economy also because of relatively easy access of variety of radiation sources, leave aside the matter of applications of radiation in medical treatments.

International Atomic Energy Agency (IAEA) has been regularly updating its maintenance activity of nuclear energy installations and those using radiation in different forms world over. In its several meetings such as SMoRE-2008 [3] held at Kharkov (Ukraine) emphasis was laid on 'Accelerator Simulations and Theoretical Modeling of Radiation Effects' with cross sections, energy spectra, and other inputs on PKA codes being the key issues. Similarly, another meeting, TM-36842 [4] on the theme 'Physics of materials under neutron and charged particle irradiations' was called with the following objectives,

1. Expert review on radiation damage in stainless steels
2. Creation of databases within the existing and planned IAEA activities involving neutron and ion irradiations
3. Initiation of a CRP on examination of advanced materials subject to high-dose neutron irradiation
4. A focused workshop or activity on dpa and a more general workshop or activity on prospects and limitations of ion irradiation to simulate neutron damage

The main recommendation of this theme meeting, TM-36842, was made as under,

> There is now a pressing need to upgrade/modify standards for materials irradiation and radiation damage produced by energetic particles and neutrons. The generally accepted reference method by which to compare irradiated test materials in different settings is DPA. In a simplistic model, dpa should be a means of normalizing experimental results independently of the irradiation source used be it either a research or power reactor, spallation source, or ion beam accelerator. Many experimental findings and numerical simulations over the past decades have shown this is not satisfactory. No simple correlation exists between DPA and microstructural radiation damage.

In light of the recommendations of the two theme meetings of a prime organization, proper modeling of radiation damage is emphasized so that the quantities like DPA and similar other quantities can be replaced to provide better standard for the sake of easy normalization of irradiated materials to different intense radiation sources including ion beams. The issue has relevance with the strong neutron/radiation sources like that used in a fusion reactor and spallation neutron source of an accelerator-driven subcritical system of energy and transmutation.

On passage of radiation through matter in the form of solid, liquid, or gas following basic changes take place,

(i) In solids, displacement of atoms is important. This generates Frenkel pairs, defects, clusters, voids, loops, and microstructures. Ionization is also an area of interest particularly in calculation of dose
(ii) Displacements and clustering, ionization, radical formation, dipole excitation, and boiling effects are important in case of liquids. Liberation of gaseous molecules such as hydrogen and helium from heavy atoms has strong effects on liquid dynamics through bubbling, etc.
(iii) Atomic excitation, ionization, and even plasma formation play more important role in gases than liquid and solids
(iv) Neutron although a neutral particle, yet, displace atoms and additionally generate atomic cascades. A gamma ray interacts with orbital electrons and the ejected electron in turn displaces atoms. In a way, this is a novel method of displacing single atoms in depth with a little fraction of atomic cascades. In case of both neutron and gamma radiation, binary collision approach (BCA) is highly useful compared to irradiation by heavy ions where atom–atom potential is relatively more

important. Also, consideration of crystallographic structure of material is important for analyzing microstructures.

Crystallographic considerations have specific effects which can be summarized in the following. For this purpose, direction of irradiation of a specific material is worth consideration.

1. **Body-Centered Cubic (BCC)**: Among the structure materials tungsten, iron, vanadium, and ferritic steels are BCC structures. Point defects in ferritic steels make it less plastic.
2. **Face-Centered Cubic (FCC)**: Copper, austenitic steels, and nickel alloys are FCC structures. Swelling is more common in austenitic steel and nickel alloys because vacancies tend to create volume clusters. In Fig. 7.3, swelling of the stainless steel under strong neutron fluence has been shown.

Swelling arising in fuel pins of BN-660 reactor is found to be due to small variations in silicon content.

3. **Hexagonal Close Packed (HCP)**: Zirconium and its alloys are HCP metals. SiC is poly-type hexagonal structure, e.g., 3C and 6H types. Radiation-induced growth is more prone in zirconium, graphite, and uranium and that can lead to dimensional changes in different directions. Although in the case of SiC for radiation resistance will be discussed later in the chapter in detail, and in Fig. 7.4 swelling and other features arising due to radiation damage are shown.

Some of the important issues of radiation damage are, for example, 'irradiation creep' which is a permanent deformation, and it depends mainly on the direction of the stress. On removal of the stress, the material does not come to the original shape and size. On irradiation,'phase transition' can also be stimulated and that can also lead to negative radiation resistance. Influence of gamma and neutron radiation as the embrittlement of steel of a reactor pressure vessel (RPV) in the 'High Flux Isotope Reactor, (HFIR)' [8–10] and corrosion of steel [11] have been observed. Helium embrittlement due to alpha emission in (n, α) reaction is also a well-known radiation effect. Similarly, hydrogen emission in (n, p) type reactions leads to bubble formation. Based on the physical behavior of gases, swelling rate of FCC austenitic steel is 1% per DPA while in case of ferritic steel [12] it is 0.2% per DPA. On one hand, Ni helps in stabilizing FCC base but on the other hand high Ni content of steels, such as PE16, becomes brittle. Obviously, reduction of Ni content of steel helps in corrosion of the steel. Conversion of Ni on impact of neutron can be studied following Eqs. (7.1)–(7.5). Alternatively, vanadium-based alloys are considered to be strong candidates for the first wall/blanket of a post-Next European Torus/International Thermonuclear Experimental Reactor, i.e., a fusion reactor, ITER [13].

$$\left.\begin{array}{l} {}^{58}_{28}\mathrm{Ni} + {}^{1}_{0}n \rightarrow {}^{59}_{28}\mathrm{Ni} \\ {}^{59}_{28}\mathrm{Ni} + {}^{1}_{0}n \rightarrow {}^{4}_{2}\mathrm{He} + {}^{56}_{26}\mathrm{Fe} \end{array}\right\} (E < 0.1\ \mathrm{MeV}) \tag{7.1}$$

$${}^{58}_{28}\mathrm{Ni} + {}^{1}_{0}n \rightarrow {}^{4}_{2}\mathrm{He} + {}^{55}_{26}\mathrm{Fe}\ (E > 0.1\ \mathrm{MeV}) \tag{7.2}$$

$${}^{59}_{28}\mathrm{Ni} + {}^{1}_{0}n \rightarrow {}^{1}_{1}\mathrm{H} + {}^{59}_{27}\mathrm{Co}\ (E < 0.1\ \mathrm{MeV}) \tag{7.3}$$

$${}^{58}_{28}\mathrm{Ni} + {}^{1}_{0}n \rightarrow {}^{1}_{1}\mathrm{H} + {}^{58}_{27}\mathrm{Co}\ (E > 0.1\ \mathrm{MeV}) \tag{7.4}$$

$$3\mathrm{Fe} + 4\mathrm{H_2O} \rightarrow \mathrm{Fe_3O_4} + 4\mathrm{H_2} + 1.7\ \mathrm{kJ} \tag{7.5}$$

7.1.2 Early Approaches of Assessment of Radiation Damage

All the old approaches of assessment of radiation damage are based on estimation of displacements per atom. It means that if DPA of an irradiated material is said to be equal to 1; then, all atoms of the irradiated material are on an average displaced one time. Here, displacement distance is an important issue as some of the displaced atoms very close to their vacated positions may recombine back to the original position because of change in some physical conditions, like

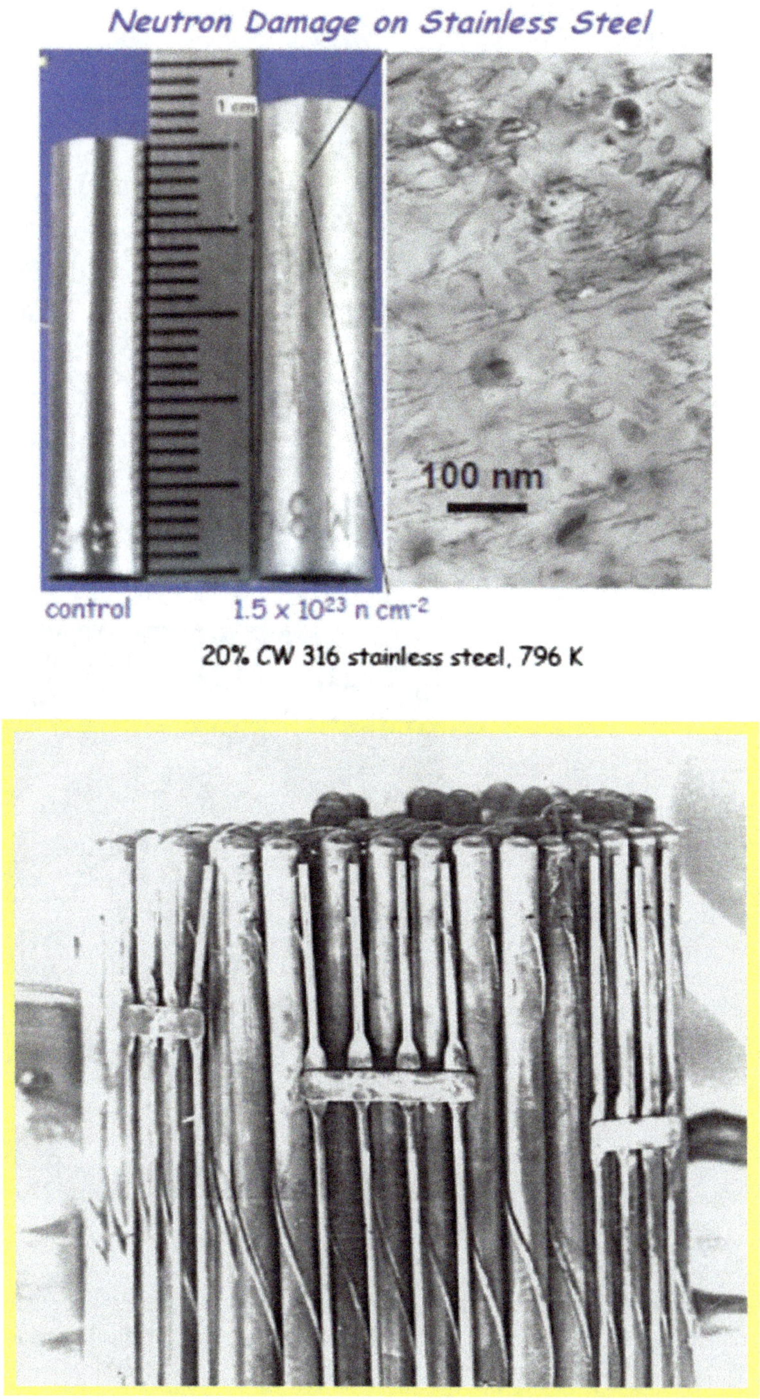

Fig. 7.3 (Up) Swelling of steel under strong irradiation [5], (Below) variation of pin length due to swelling in EI-847 fuel pins irradiated in BN-600 [6]

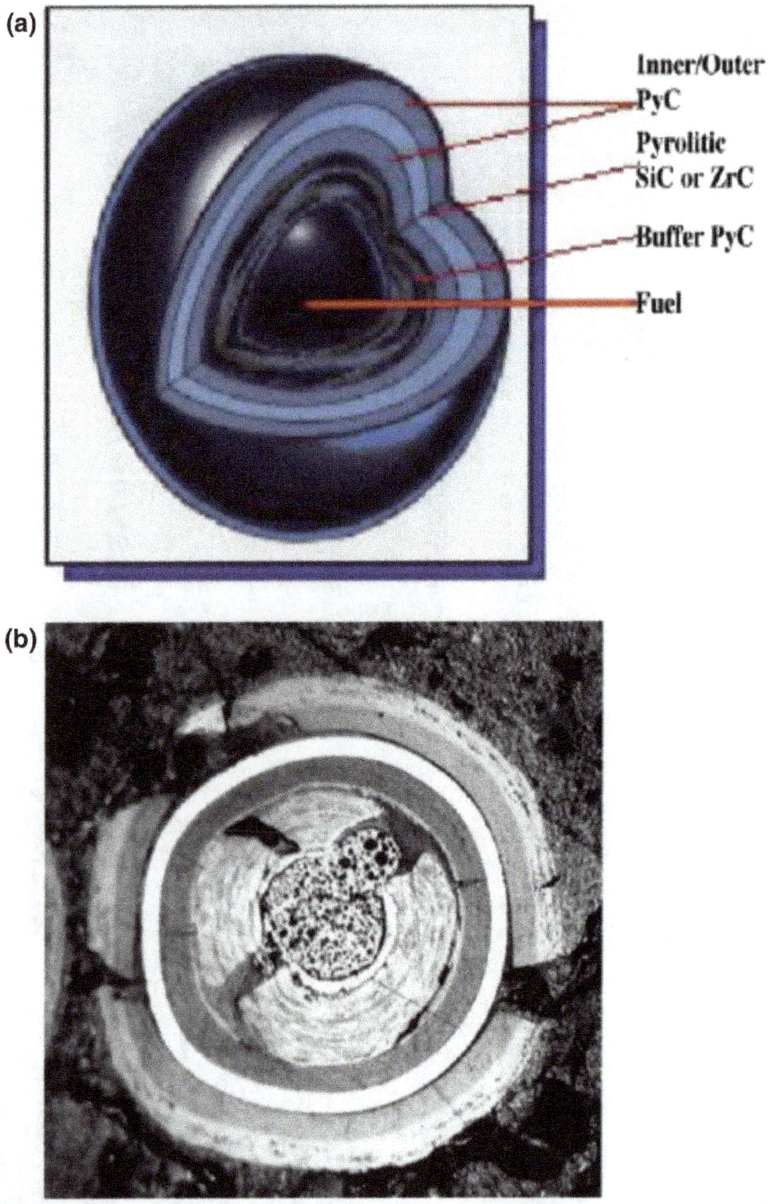

Fig. 7.4 (Up) A fuel cell for an advanced energy system before irradiation with SiC coating and (down) fuel cell after application in a reactor [7]

temperature. The calculation of DPA has been popular in case of structure and shielding materials of a reactor when exposed to neutron, gamma and fission products. The fission products lose energy very fast by the way of ionization. In case of fast reactors, fusion and ADSS kind of reactors displacements may also occur due to the following processes,

(i) Energy transfer during an elastic or an inelastic scattering of a neutron,
(ii) Recoil of a nucleus on emission of an energetic particle, usually subsequent to a capture reaction of a neutron or (n, α), (n, p), (n, d), (n, γ) and other reactions,
(iii) Energy transfer by another knocked off or recoil atom, or

(iv) Energy transfer from a secondary (emitted) particle.

In fact, the produced particles generate another series of interactions and recoils. The produced particles may also stay inside at the position of its stopping in a vacant position or close to the atom of the material. All such processes contribute toward damage of material. Rate of radiation damage by a neutron of energy, E per unit time and per unit volume is given by,

$$D(E) = N\sigma_d E\varphi(E) \tag{7.6}$$

Where N is the atom density, φ is the neutron flux (no./cm^2/s), and σ_d is the microscopic damage cross section (cm^2). As mentioned above, there can be a number of ways of producing displaced atoms by elastic scattering or recoils produced in primary collisions like (n, α), (n, p), (n, d), (n, γ), or even secondary knocked-on atom (SKA) due to primary knocked-on atom (PKA). Total displacement cross section can be written as summation over all such processes,

$$\sigma_d(E) = \sum_n \sigma_{dn}(E) \tag{7.7}$$

Displacement cross section due to any primary type reaction can be calculated from

$$\sigma_{dn}(E) = \int_{E_d}^{E_T\max} \sigma_n(E) P_n(E, E_T) \nu(E_T) \mathrm{d}E_T \tag{7.8}$$

Here, P_n is the probability of nth type interaction with transferred energy E_T to a PKA out of the incident, E energy. The $\nu(E_T)$ is the effective number of displaced atoms due to transfer of E_T energy. Specifically, at low energy of neutrons elastic scattering is the dominant interaction.

7.1.2.1 Estimation of $\nu(E_T)$

Dienes and Vineyard [14] have systematically described how different models estimate $\nu(E_T)$ corresponding to transferred energy, E_T by a projectile to a PKA by calculating E_T from kinetics of a collision. For isolation of an atom from the bonding of other neighboring atoms of a lattice minimum threshold energy, E_d is needed to be transferred to the atom and this varies from material to material. In Table 4.3, values of E_d for a large number of elements have been given. According to one of the initial models, Kinchen and Peas [15],

$$\nu(E_T) = 0 \quad 0 < E_T < E_d \tag{7.9}$$

$$\nu(E_T) = 1 \quad E_d < E_T < 2E_d \tag{7.10}$$

$$\nu(E_T) = E_T/(2E_d) \quad 2E_d < E_T < E_{T1} \tag{7.11}$$

This shows that when the transferred energy, E_T rises beyond 2E$_\mathrm{d}$, then $\nu(E_T)$ rises linearly with E_T up to E_{T1} and beyond this it is saturated,

$$\nu(E_T) = E_{T1}/2E_d \quad E_T > E_{T1} \tag{7.12}$$

As a matter of mechanism, hardcore scattering is assumed up to E_{T1} and all atoms receiving energy $<E_d$ return back to the original position. Later, Nelson model [16] corrects the formulation for the realistic scattering in place of hardcore approximation and the defect recombination. Torrens and Robinson [17] proposed a modified Kinchen-Peas model by applying electron excitation and redefined E_{T1} by $E_T - Q$ where Q is the energy lost in cascades. Also, they identified displacement efficiency factor to be $k \sim 0.8$. In the high E_T—region,

$$\begin{aligned} \nu(E_T) &= k(E_T - Q)/2E_d; \\ E'_T &= E_T - Q > 2E_d/k \end{aligned} \tag{7.13}$$

Later, the modified Kinchen-Peas model has been incorporated in NRT [18] model where energy invested in displacements is calculated according to the Lindhard et al. [19] approach for the whole range of energy. The NRT formula is widely adopted in calculations of radiation damage,

$$\nu(E_T) = kE'_T/2E_d \tag{7.14}$$

with

$$E'_T = E_T/[1 + k_L g(\varepsilon)],$$
$$g(\varepsilon) = 3.4008\varepsilon^{1/6} + 0.40244\varepsilon^{3/4} + \varepsilon,$$
$$k_L = 0.1337 Z_1^{1/6}(Z_1/A_1)^{1/2},$$
$$\varepsilon = [A_2 E_T/(A_1 + A_2)]\left[a/Z_1 Z_2 e^2\right],$$
$$a = (9\pi^2/128)^{1/3} a_0 \left[Z_1^{2/3} + Z_2^{2/3}\right]^{-1/2} \text{ and}$$

Here, a_0 is the Bohr's radius and suffix 1 and 2 are used for the projectile and target mass number (Z) and atomic mass (A) respectively and ε is a dimensionless energy. In a detailed discussion, Gopalakrishnan [20] has pointed out that (i) in the model E_d is taken to be 40 eV for all elements of the steel (ii) as the model depends mainly on the energy of PKA and in different interactions it will be different and for complicated inelastic interactions recoil energies will also involve angular anisotropy and (iii) the model assumes that all small displacements or the defects produced on irradiation will recombine back; therefore, their estimation is not considered in the model.

7.1.3 Monte Carlo Simulation Approach of JA-IPU Code

In the modern time with the availability of high computation power, it is imperative to develop a detailed code based on Monte Carlo simulations of the entire process of transport of radiation or particle like neutron, proton and heavy ion or even a photon through a complex material of several elements with the help of binary collision approximation involving all kinds of physical effects like ionization loss, excitations, displacement distances of recoils hence separation of displaced Frenkel pair (interstitial and vacancy) or the defects. This category of codes of radiation damage can accommodate the situations like escape out of an ejected ion or atom from the irradiated sample. In the following part of the chapter, details of the JA-IPU code with its two versions one for neutron [21] and other for gamma radiation [22] will be discussed. In the code, chemical effects of irradiation like radical formation and combination are not considered. Several other codes involving dynamical behavior of colliding partners, time evolution of the event, velocity and temperature dependence of potentials, etc., have also been developed [23–25]. They are highly perspective as well as attractive from the point of exhausting observable physical and chemical aspects such as evolution of event on accommodating large number of variables and parameters. Such complications normally slow down the calculations. To discuss the dynamical modeling (DM) of radiation damage, a big volume of the text will be required hence it will be skipped. In the following discussion of JA-IPU code based on binary collision approach is presented in detail, and this also includes comparison of some of the results of the dynamical models.

7.1.3.1 Binary Collision Approach (BCA)

Binary collision of two particles is treated like collision of two billiard balls without involving any complicated field in between the two particles. Particularly, a neutron collision at low energy may safely be identified as a binary collision because elastic collision cross section is much higher than inelastic cross section including the absorption cross section. At high energies, neutron wavelength is shorter up to several tens of MeV energy, intra-particle approach will dominate and it may no longer be treated as a pure binary collision because of active involvement of many individuals for the sake of outcome of a collision. A collective behavior plays certain amount of role. Thus, at low energies most of the older models [17, 26] have adopted binary collision approach (BCA), for example, TRIM [27, 28] adopts BCA and it makes use of Monte Carlo techniques to describe the trajectory of the incident particle and the damage created in amorphous solids. Similarly, MARLOWE code [29, 30] also uses BCA and the code simulates the atomic collisions in crystalline targets. Incident projectiles can be an external beam or an internal source site from where a radiation is introduced in rest of the medium. The code

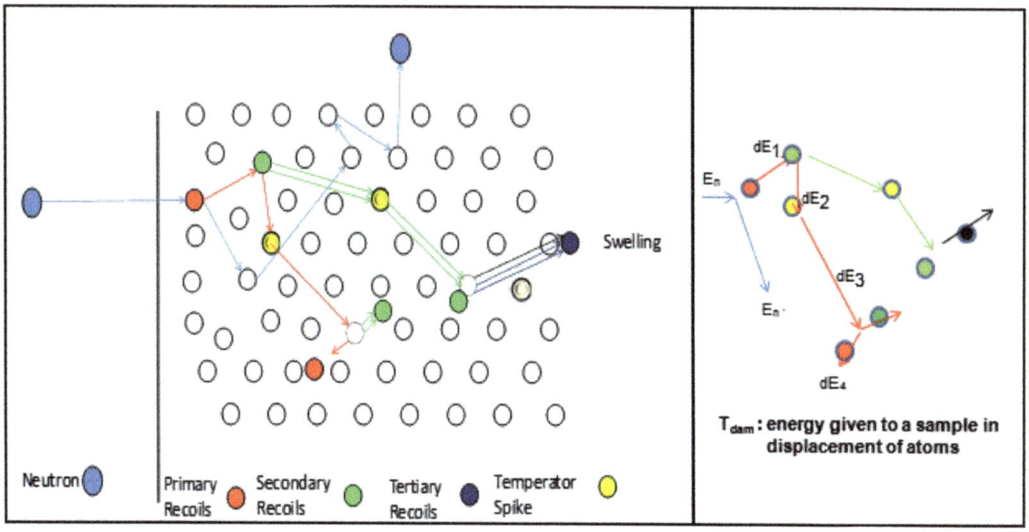

Fig. 7.5 Transport of a neutron in material and displacement of atoms. Both loss of KE and escape out are shown along with excited atom marked as spike

records tracks of a binary collision and the displaced atoms from their lattice sites until they either leave the medium or fall below the selected kinetic energy, i.e., displacement energy E_d. Calculations of damage of the medium have also been included in the code.

Development of the JA-IPU code is described at length in references [21, 22]. Accordingly, success of a code depends on procedure of random number generation and the number of generated histories of an event. Transport of a particle is assumed to be a straight line trajectory of length, l and on scattering from a target atom elastically it is diverted at polar angle, θ, and azimuthal angle, ϕ. Probability of path that a neutron for example, interacts in between the distance l and $l + \delta l$ is given by [31],

$$P(l)\mathrm{d}l = \Sigma_{\mathrm{T}}\delta l e^{-\Sigma_T l} \tag{7.15}$$

Here, Σ_{T} is the total macroscopic cross section which can be expressed as,

$$\Sigma_{\mathrm{T}} = \Sigma_{\mathrm{S}} + \Sigma_{\mathrm{A}} = \Sigma_{\mathrm{EL}} + \Sigma_{\mathrm{IN}} + \Sigma_{\mathrm{A}} \tag{7.16}$$

Here, Σ_{S} is scattering, Σ_{A} is absorption, Σ_{EL} is elastic, and Σ_{IN} is inelastic macroscopic cross sections related to the interaction. For the reason that an interaction will be independent of the previous one, hence random, thus one can derive a relationship between random variable, R and interaction distance, l as the following,

$$l = -\frac{1}{\Sigma_{\mathrm{T}}}\ln(R) \tag{7.17}$$

and the probability of interaction,

$$P(l) = 1 - \mathrm{e}^{-\Sigma l} \tag{7.18}$$

After the interaction, when both the particle and recoil atom move then their new directions are decided by assigning new set of (θ, ϕ) coordinates with the help of another sets of two random numbers [31] independent to each other.

Using these simple geometrical aspects, one can assign directions to the recoils and the angular relationship between the incoming projectile and the recoiled atoms using kinetics of elastic collision.

Independent two cascading of primary projectile and the recoiled secondary atom are depicted in Fig. 7.5. The knocked-on atom by the projectile is designated as a PKA which later on generates its own cascade as shown in Fig. 7.6. The ongoing scattered projectile or the PKA may

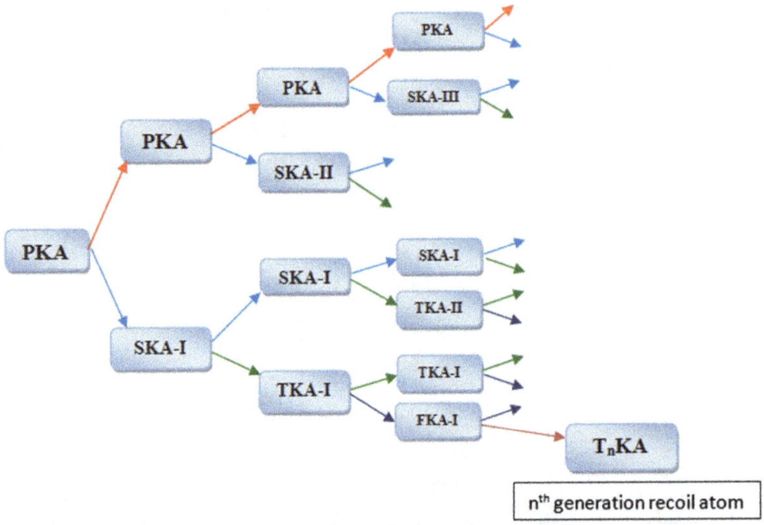

Fig. 7.6 Atom–atom collision cascade generated by a primary knocked-on atom (PKA)

have another chance to produce next set of the atomic cascades. Thus, two kinds of cascades are generated, one due to projectile particle and other due to a PKA.

When a particle moves through a matter having either of the three, elastic, inelastic and absorption kind of interactions then they can be sorted out by applying following conditions,

$$\begin{array}{ll} \text{If } 0 \leq R \leq \frac{\Sigma_{el}}{\Sigma_T}, & \text{interaction is elastic.} \\ \text{If } \frac{\Sigma_{el}}{\Sigma_T} \leq R \leq \frac{\Sigma_{in}}{\Sigma_T}, & \text{interaction is inelastic.} \\ \text{If } \frac{\Sigma_{in}}{\Sigma_T} \leq R \leq 1, & \text{particle is absorbed.} \end{array}$$

In such simulation obviously, the material is supposed to be uniform density of single element. In case of composite target of more than one element, e.g., the case of SiC, selection of the target atom facing a collision is done through the sampling of random number, R, if $0 < R \leq \frac{\Sigma_{T,Si}}{\Sigma_T}$ for interaction to take place with Si atom and it will be interacting with C atom if $\frac{\Sigma_{T,C}}{\Sigma_T} < R \leq 1$. Here, $\Sigma_T = \Sigma_{T,Si} + \Sigma_{T,C}$. In this text, Monte Carlo simulation of neutron irradiation of three materials, e.g., Ni, Nb, and SiC are considered.

Monte Carlo Simulation of Neutron Irradiation

For the simulations of neutron cascade $n + A$, cross section data are taken from evaluated ENDF/B-VII.0 library and plots for (a) n + Ni (b) n + Nb (c) n + Si and (d) n + C collisions are given in Fig. 7.7. It can be noted that at energies above the thermal energy up to a few MeV elastic collision cross sections in case of the four targets are much higher than inelastic cross sections barring a few energies around 10 eV in case of Ni and Nb. At such low energies, a target nucleus can only be excited and its recoil or explosion or evaporation will not take place being below E_d of the material. On the contrary, in case of fissile elements there will be a high possibility of nuclear fission.

Atom + Atom Collision Cross Section

When a neutron interacts with an atom, it transfers energy, T to the atom which may result in the displacement of atom from its original lattice site or it goes in excitation state. The recoiled atom is referred as PKA (Primary Knocked-on Atom). In fact, nucleus of the atom is assumed to be displaced along with its electron cloud. If the energy transferred to the atom is much more than the threshold energy E_d ($T > 2E_d$), the recoiled atom may further interact with another atom(s).

Mean free path, L of the displacement of atom is inversely proportional to the interaction cross section, $\sigma(E)$, i.e., higher the cross section, lesser is the m. f. p. length, where E is the PKA energy, for example.

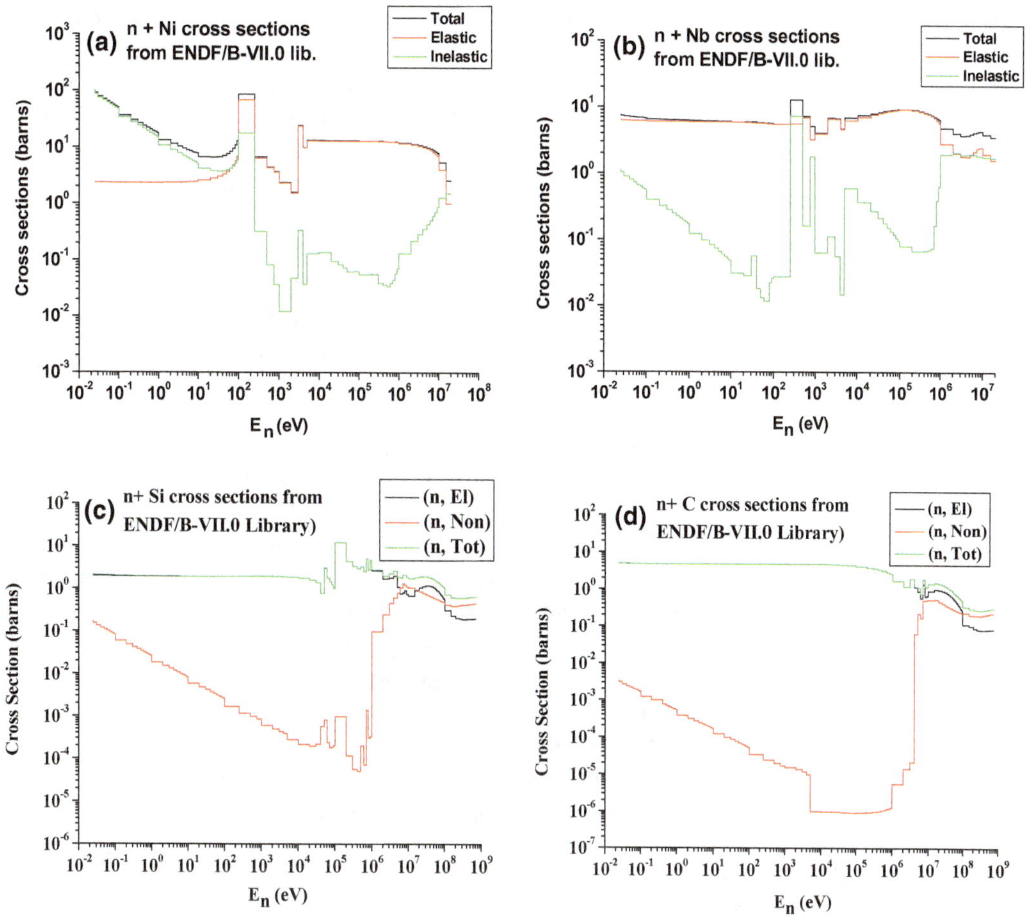

Fig. 7.7 Elastic, inelastic, and total evaluated neutron cross sections of **a** Ni, **b** Nb, **c** Si, and **d** C targets from the ENDF/B-VII.0 library

For the interaction cross section, σ_{int} at projectile energy, E to transfer energy, T to the target, the code uses the algorithm of IOTA code [32],

$$\begin{aligned}\sigma_{\text{int}}(E) &= \sum\nolimits_{i=1}^{N} w_i \sigma_{\text{int}}^{i}(E) \\ &= \sum\nolimits_{i=1}^{N} w_i \int_{E_d^i}^{T_{\max}^i} \mathrm{d}\sigma_i(E, T)\end{aligned} \tag{7.19}$$

Here,

w_i Atomic fraction of the ith component of the target material

E_d^i Effective threshold displacement energy of the ith component

$T_{\max}^i$ Maximal energy transferred from the incident ion to PKA

Differential cross section $\mathrm{d}\sigma(E, T)$ is calculated as follows,

$$\mathrm{d}\sigma_i(E, T) = \pi a^2 f\left(t^{1/2}\right) \frac{\mathrm{d}t}{t^{3/2}} \tag{7.20}$$

where the function $f(t^{1/2})$ is calculated from the following equation,

$$f\left(t^{1/2}\right) = \lambda t^{1/2-m}\left[1 + \left(2\lambda t^{1-m}\right)^{q}\right]^{-1/q} \tag{7.21}$$

In the JA-IPU code, values of λ, m and q parameters for IDSDT = 0 are taken from Table 1 of the IOTA manual Ref. [33]. Here, $\lambda = 1.309$, $m = 1/3$, $q = 2/3$ and

$$t = \frac{\varepsilon T}{\alpha^2 \varepsilon_0} \tag{7.23}$$

$$\varepsilon = \frac{E}{\varepsilon_0} = \frac{aA_2E}{Z_1Z_2e^2(A_1 + A_2)} \tag{7.24}$$

$$\alpha = \frac{4A_1A_2}{(A_1 + A_2)^2} \tag{7.25}$$

where Z_1, A_1, and Z_2, A_2 are the atomic number and mass number of the projectile and the target material respectively. Threshold displacement energies are taken as [33]. Interaction cross sections of ^{63}Ni + ^{63}Ni, ^{93}Nb + ^{93}Nb, ^{12}C + ^{12}C, ^{12}C + ^{28}Si, ^{28}Si + ^{12}C, and ^{28}Si + ^{28}Si have been calculated using the IOTA code and shown in Fig. 7.8a–f, respectively.

On irradiation when an atom is displaced from its position (in a crystal it is a lattice position) then a Frenkel pair is formed by the vacancy and the displaced atom. If displacement is small, i.e., $r_d < 0.75$ Å [34], then it is assumed that there is a definite chance that the atom recombines back at the original position whenever getting suitable amount of kinetic energy. Primarily, because energy is invested in the small displacement so we count energy and call it as 'defect'. Thus, 'displacements' and 'defects' are produced on irradiation along with the hot spikes which are the excited atoms. A frequent cascading of a particle may produce clusters and loops as well as a cluster of excited atoms called 'spikes'.

Irradiation by Gamma

Damage of a medium by gamma rays takes place through electrons produced in Compton, photoelectric and pair production processes. Its Monte Carlo simulation is more cumbersome than by a neutron because of the additional orbital structure of atom and interaction of gamma with electron. For that purpose, in JA-IPU code, lattice structure of a material is generated. Point cross sections of the three processes are taken from the ENDF/B-VII.0 library [35]. In the code, cross sections have been parameterized in different gamma energy regions before making use in the code for a faster computation. On interaction, the ejected electron loses energy by way of ionization, Brehmsstrahlung, and finally on colliding with an atom. Elastic collision of electron with atom is considered as a billiard ball collision and collision cross sections are taken from a compilation by Mayol et al. [36]. For details, reader can refer to [22]. After an atom is displaced the atomic cascade is handled as in the case of neutron using the IOTA code. In Fig. 7.9, frequency of displacement distance of an atom from its lattice position shown as the simulated path lengths is plotted in case of a 50 keV vanadium (V) recoil colliding with vanadium in the medium. This generates cascades through the material and it can be seen that the total number of path lengths < 0.75 Å which represent production of defects is comparable with the path lengths > 0.75 Å.

N_d Versus T_{dam} Regularity

Damage energy is defined as the energy that is invested in changing the pristine structure of the material. When an atom displaced from its original position getting energy more than E_d then it moves in the medium and loses its kinetic energy by ionization as well as displacing another atom of the medium. This can be a 'displacement' or a 'defect'. It is also possible that the particle escapes out of the medium with certain nonzero energy, E_{esc}. Thus, $T_{dam} = \mathrm{KE} - E_{ioni} - E_{esc}$. In case atom gets much larger energy than E_d, then it may further produce next generations of cascades. In Fig. 7.10, results of MC simulation of N_d and T_{dam} are plotted for the Ni and Nb materials for the spallation neutron spectra produced by a 660 MeV proton colliding with a thick Pb target [37]. The two quantities N_d and T_{dam} (eV) are calculated for each neutron per incident proton, (n/p). Similar plots have been worked out in the case of SiC irradiated by neutron spectra of AmBe and spallation neutrons [38] separately. In all cases of different materials, e.g., Nb, Ni, and SiC there is a linear behavior.

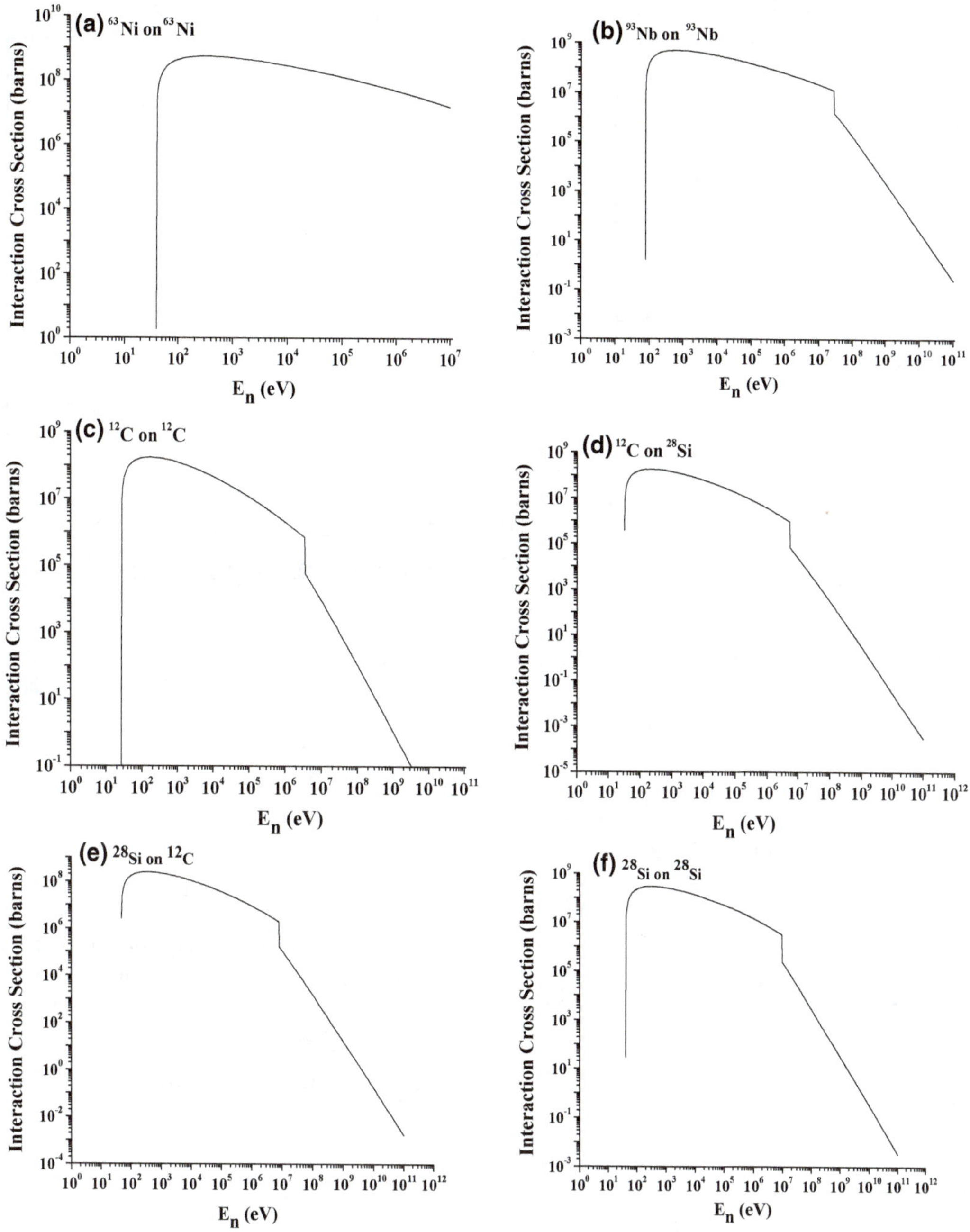

Fig. 7.8 Interaction cross sections of **a** ^{63}Ni + ^{63}Ni, **b** ^{93}Nb + ^{93}Nb, **c** ^{12}C + ^{12}C, **d** ^{12}C + ^{28}Si, **e** ^{28}Si + ^{12}C, and **f** ^{28}Si + ^{28}Si calculated by the JA-IPU code

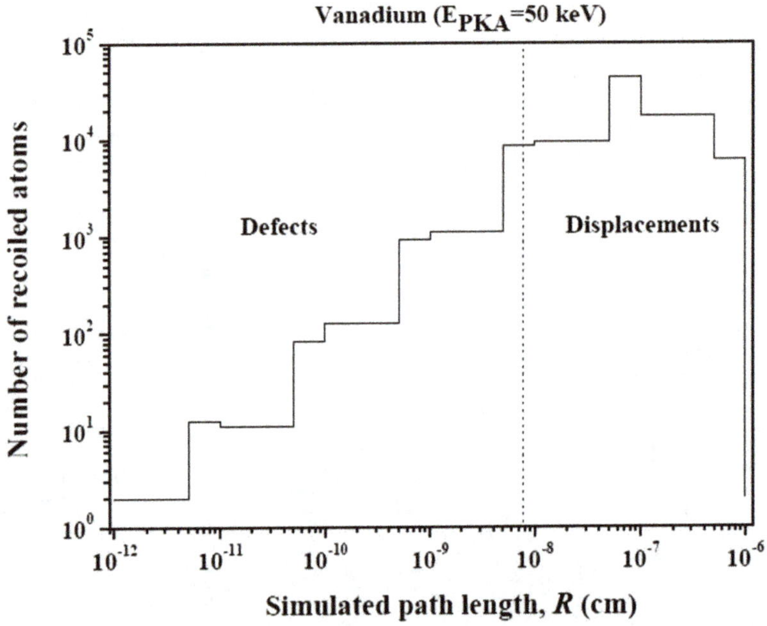

Fig. 7.9 Number of defects and displacements in transport of 50 keV vanadium ion through the vanadium medium

Similar linear behavior between N_d and T_{dam} is seen in case of incident beam of gamma on the Fe sample shown in Fig. 7.11 [38].

7.1.4 A Phenomenological Approach of Radiation Damage

Monte Carlo simulations of a large number of materials performed for irradiation by different neutron spectra and gamma, a common linear behavior between N_d and T_{dam} is noticed as shown in Figs. 7.10 and 7.11. In Table 7.1, values of slope k (eV^{-1}) are calculated from the basic plots given in the two figures between N_d and T_{dam} measured in units of eV/n in case of AmBe n-source and eV/n/p in case of spallation neutrons.

In case of pure iron irradiated by gamma spectrum of Co^{60} source [39] value of slope, $k = 0.0234 \pm 0.0004\ \mathrm{eV}^{-1}$ for the absorbed doses varying between 0.003 and 0.319 Gy. In irradiation of iron by the Co^{60} gamma spectrum, production of single displacements [39] has been observed which may not be possible in irradiation by neutron and heavy ion beams.

A phenomenological approach is developed by Kumar et al. [39] to correlate the displaced atoms of the irradiated material with the physical observables followed from the said regularity between N_d and T_{dam},

$$\mathrm{d}N_d \propto \mathrm{d}T_{\mathrm{dam}} \tag{7.26}$$

$$\mathrm{d}N_d = k\mathrm{d}T_{\mathrm{dam}}$$

with k being the proportionality constant and it is spectrum dependent. Defining damage volume, $V = (N_d/n_d)$ and $n_d = f(V, T_{\mathrm{dam}})$,

It leads to [39] the following relation between change in N_d and change in volume, $\left(\frac{\mathrm{d}V}{V}\right)$ arising due to $\mathrm{d}T_{\mathrm{dam}}$,

$$\mathrm{d}N_d = \left(\frac{k}{V}\right)\mathrm{d}T_{\mathrm{dam}} - n_d\left(\frac{\mathrm{d}V}{V}\right) \tag{7.27}$$

Following Brinkman [40],

$$\frac{\rho - \rho_0}{\rho_D - \rho_0} = \frac{\mathrm{d}V}{V} \tag{7.28}$$

ρ_0 and ρ_D correspond to the resistivity of pristine and highly disordered state of a sample,

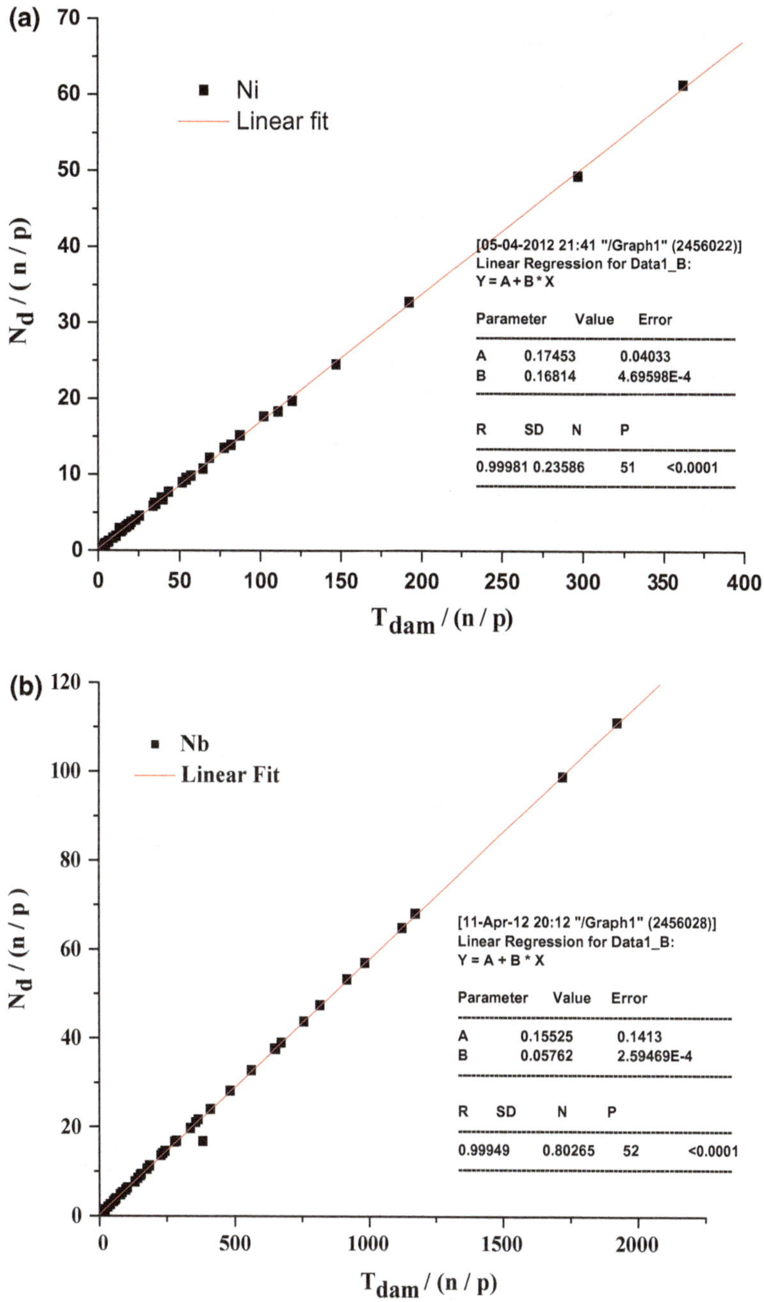

Fig. 7.10 MC simulation results of N_d (*n*/*p*) versus T_{dam} (eV/*n*/*p*) plotted for spallation neutron spectrum in case of (**a**) Ni and (**b**) N_b samples [37]

respectively. In the situation of non-availability of data of resistivity, ρ_D we assume that $\rho_D = x\rho_0$. Here, parameter $x > 0$ corresponds to a disordered state of irradiated material compared to pristine and it is also spectrum dependent because different particle spectra may reach to a different disorder. For example, $\mathrm{d}V = 0$ for no change in volume and this leads to $\rho = \rho_0$. Also, $\mathrm{d}V = V$ corresponds to $\rho = \rho_D$ in case of relation (7.28). In that sense, much higher value of x than

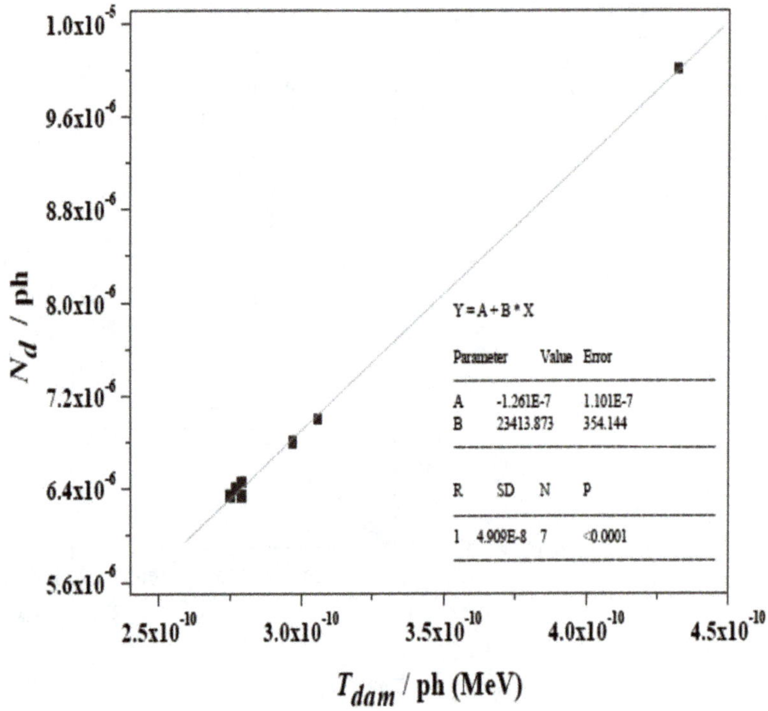

Fig. 7.11 N_d (ph) versus T_{dam} (MeV/ph) plot for pure iron sample irradiated with different doses of Co^{60} beam [39]

Table 7.1 Sizes of SiC, Nb, and Ni samples and slope, 'k' (eV^{-1}) deduced from the N_d and T_{dam} plots from the data of MC simulations

Dimensions of the samples		Slope 'k' of T_{dam} versus N_d plot	
		AmBe n-spectrum	Spallation n-spectrum
SiC	$1 \times 1 \times 0.0314\ \text{cm}^3$	0.00062 ± 0.00003	0.05992 ± 0.00066
Nb	$2.4 \times 0.8 \times 0.024\ \text{cm}^3$	$0.00138 \pm 0.\ 00001$	0.05762 ± 0.00026
Ni	$2.5 \times 1.5 \times 0.002\ \text{cm}^3$	0.875	0.16814 ± 0.00045

zero may lead to more interesting results. Thus, relation (7.28) can be written as,

$$\frac{d\rho}{(x\rho_0 - \rho_0)} = \frac{dV}{V} \tag{7.29}$$

which leads to

$$d\rho = \rho_0(x-1)\left(\frac{k}{N_d}\right)dT_{\text{dam}} \tag{7.30}$$

where both 'k' and 'x' are spectrum dependent. It may be mentioned that small deformations lead to reduction of free volume between the atoms.

Taking $k = \frac{dN_d}{dT_{\text{dam}}}$, relation (7.30) can be written as,

$$(\rho - \rho_0) = \rho_0(x-1)\ln N_d \tag{7.31}$$

The relation (7.31) is applicable for irradiations where single displacements take place, for example, in case of incident gamma spectra, and it has been established using the positron annihilation spectroscopy [41]. Also, it shows rise of resistivity on increase of displaced atoms. The parameter x has prospects in designing of devices by ion beam irradiation.

7.1.5 Dynamic Model Approach and JA-IPU

As mentioned earlier, dynamic modeling of radiation damage is a much-involved process than simpler BCA approach. This has however capability of handling the natural processes, for example, dynamics of proteins and DNA on a timescale ranging from 10^{-9} to 10^{-6} s. To make statistically valid conclusions from the simulations, simulated time period should match the kinetics of the natural process. This is done through fine-tuning of speed of calculation and natural timings of a process. In real sense, it is 'a state-of-the-art' technique because of involvement of large number of variables and by bringing computation close to the speed of happening of an event. The technique is useful at level of small atomic energies. Using a dynamical model, (DM) events at the atomic scale say up to 10^{-11} s can be successfully simulated but not at lower timescales like femtosecond (10^{-15} s). Transmutation of the inner wall of the fusion reactor leads to fast degradation of the wall material at the timescale of dynamical modeling. Kinetic Monte Carlo (KMC) is a more effective technique, which simulates events randomly with probabilities according to the corresponding event rates [42]. It provides [43] self-estimate of the time step as the simulation proceeds and is the most powerful approach available for making a dynamical prediction of mesoscale events. In case of radiation damage of a static medium, the two approaches are ineffective because of no consideration of movements of orbital electrons and vibrational and collective modes of oscillation of an atom.

Alonso et al. [25] have calculated the results by coupling DM with Kinetic Monte Carlo (KMC) simulations of damage production efficiency, η of a code in case of 20 keV vanadium as PKA traversing the vanadium medium. Damage production efficiency is defined as,

$$\eta = N_d / N'_{\mathrm{NRT}} \tag{7.32}$$

$$\text{with } N_{\mathrm{NRT}} = 0.8 E_{\mathrm{PKA}} / (2E_d) \tag{7.33}$$

Using the JA-IPU code damage production efficiency, η is calculated [44] for different energy of PKAs as shown in Fig. 7.12 considering electronic energy loss (EEL) and without considering EEL. It can be noted that DM approach gives $\eta \sim 26\%$ comparable with $\sim 28\%$ from the JA-IPU code.

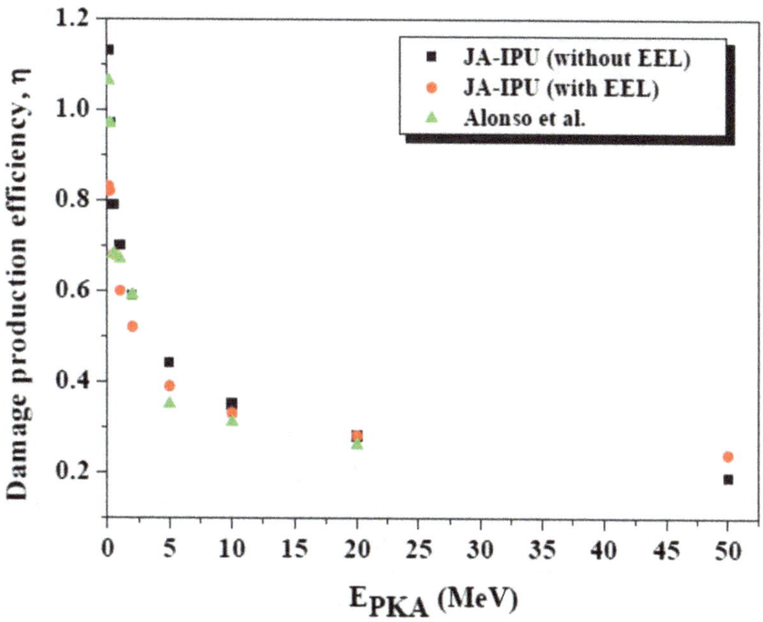

Fig. 7.12 Damage production efficiency plotted for different energy PKAs for vanadium [44]

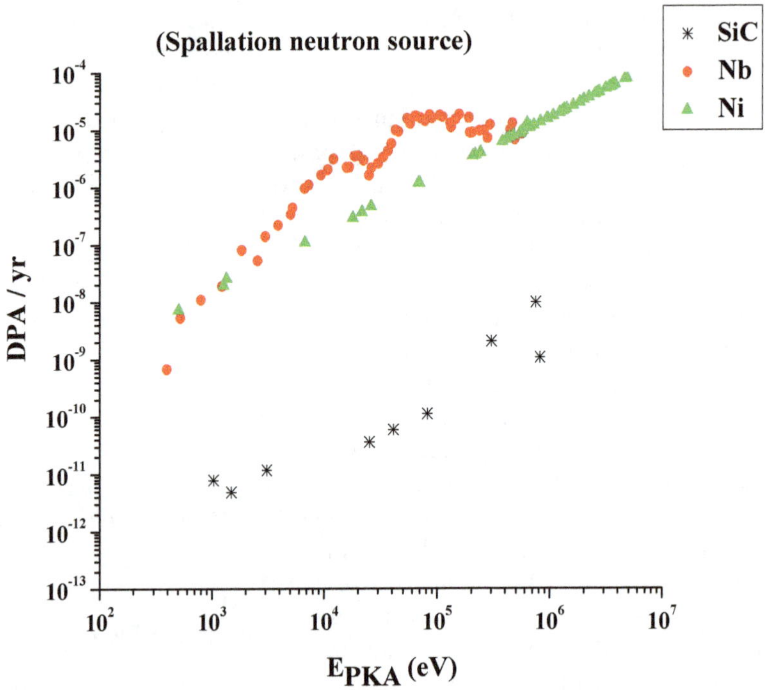

Fig. 7.13 DPA/year plotted with respect to E_{PKA} for the irradiation by the spallation neutron spectra [37]

Table 7.2 Enhanced values of resistivity $(\rho - \rho_0)$ and $\langle \sigma T_{dam} \rangle$ b keV values for SiC and Nb samples [45]

Sample	ρ_0 (Ω cm)	ρ (Ω cm)	$\rho - \rho_0$ (Ω cm)	$\langle \sigma T_{dam} \rangle$ (b keV)
SiC (spall)	2.45E−4	2.98E−4	0.53E−4	3.33E+10
Nb (AmBe)	4.40E−6	1.491E−4	1.447E−4	6.33E+12
Nb (spall)	4.40E−6	2.664E−4	2.62E−4	2.44E+15

As noticed, the phenomenological approach developed as above opens ways for searching of suitable variables for answering the questions raised on the DPA and similar other quantities used to monitor the radiation damage. On calculating DPA using number of displaced atoms including defects, N_d from the JA-IPU code, and the E_{PKA} of an irradiation by the spallation neutron spectra in Fig. 7.13, DPA/year is plotted with E_{PKA} and it can be seen that the three materials, Ni, Nb, and SiC follow different variations. In case of the three materials average, DPA/year is 8.62E+1 for Ni, 2.828E−2 for Nb, and 6.75E−6 in case of SiC, respectively. In case of a softer AmBe neutron spectrum, DPA/year is 1.521E−7 and 1.03E−7 for Ni and Nb, respectively, showing spectrum dependence of DPA/year.

In Table 7.2, data of change of resistivity $(\rho - \rho_0)$ and 'average damage energy cross section' $\langle \sigma T_{dam} \rangle$ is given for the SiC and Nb for the spallation neutron spectra additionally for Nb for the AmBe neutron spectra also. In Fig. 7.14 (a), $(\rho - \rho_0)$ versus $\langle \sigma T_{dam} \rangle$ plot is shown and it is qualitatively compared with $(\rho - \rho_0)$ versus $(x - 1)\ln N_d$ plot of Fig. 7.14b. Data of resistivity ρ is experimentally measured and given in the two plots and presented in Table 7.2. Values of N_d as well as $\langle \sigma T_{dam} \rangle$ are taken from the MC simulations using the JA-IPU code.

From the plots of Fig. 7.14, it may be inferred that the quantity like $\langle \sigma T_{dam} \rangle$ may be treated unique from the point of both target and different neutron spectra and it can probably replace earlier used quantity like DPA. More experiments may be planned with high energy gamma, ion beams to produce data of $\langle \sigma T_{dam} \rangle$ and measurable physical quantity like change in resistivity to validate the aforesaid observation.

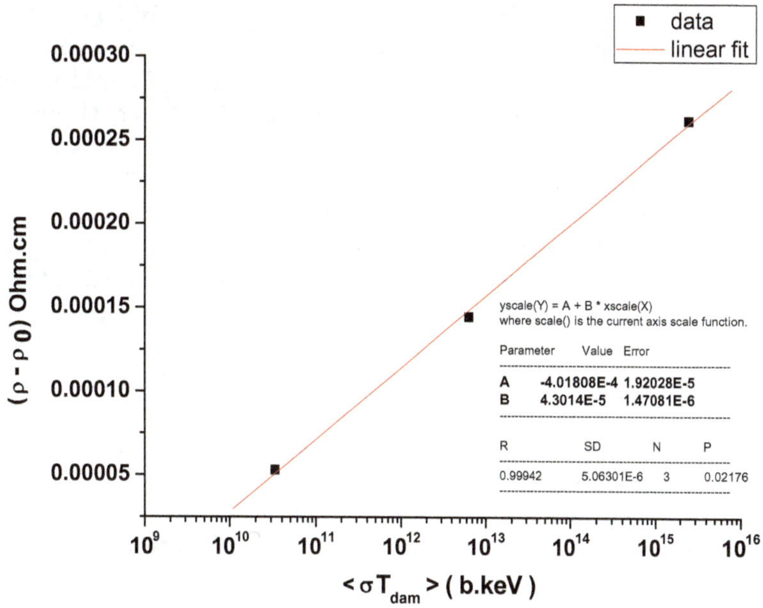

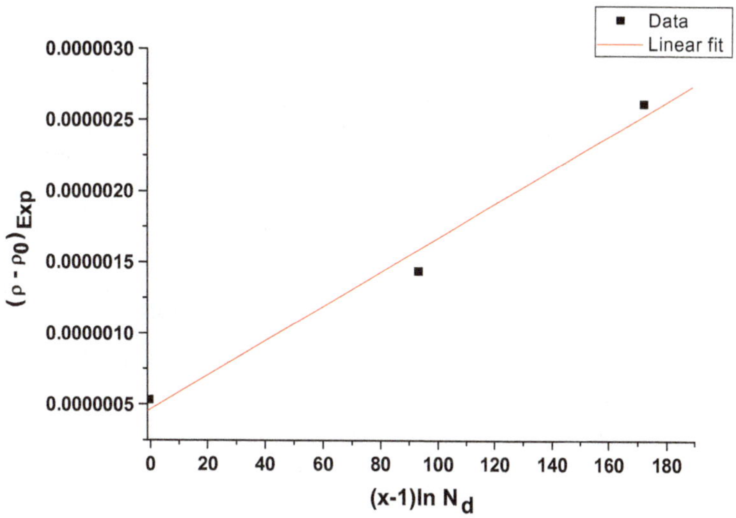

Fig. 7.14 (Up) $(\rho - \rho_0)$ versus $\langle \sigma T_{\mathrm{dam}} \rangle$ plot and (below) $(\rho - \rho_0)$ versus $(x - 1)$ $\ln N_d$ plot for the two spallation and AmBe neutron spectra as already discussed [37, 38]

In a study of radiation damage of HFIR, high rate of embrittlement of steel of high-pressure reactor by high energy flux of both gamma and neutrons was noticed [8, 46] and it was stressed that there is need of MC study of damage by gamma radiation because it can be understood that the neutron flux will be highly moderated at 30 cm passage than the gamma in the reactor. Using the JA-IPU code, radiation damage by gamma was estimated [47] and found it to be comparable with the observed results.

7.1.6 Radiation Resistant Materials

Emphasis is being laid on research and development of the radiation resistant materials (RR) both for the fuel cells, structure materials, shielding materials of gen. IV reactors including ADSS and fusion reactors. Heating and diffusion of radioactive material in cassettes and canisters used in reposition of the HLW is another serious concern and it demands development of RR materials with attention on the said problems. In case of reposition, Ewing et al. [48, 49] have suggested that planning of development of the radiation resistant material needs to incorporate understanding of ion-matter interactions from the point of immobilization of high-level radioactive materials. Ceramics materials such as pyrochlore $A_2B_2O_7$, zirconolite $CaZrTi_2O_7$, zircon $ZrSiO_4$, monazite (La, Ce, Nd) PO_4, and other complex oxides having high waste loading capacity, better thermal, mechanical, natural analogous and desired chemical properties are proposed as potential materials for the immobilization of high-level nuclear waste and waste form of ADSS. Among these ceramic materials, pyrochlore structure shows properties of inert fuel matrix for the radionuclides. In addition, some of the pyrochlores also have potential to be used as burnable neutron poisons in a nuclear reactor. In particular, isometric pyrochlores with the $A_2B_2O_7$ stoichiometry display capability to incorporate actinides (e.g., U, Np, Th, Pu, Am, and Cm) at the A-site, particularly Pu [50]. Compounds with strong ionic characteristics, such as zirconate ($A_2Zr_2O_7$), are generally considered to be more radiation resistant than covalent stannate ($A_2Sn_2O_7$). The in-situ TEM studies during ion irradiation showed that critical dose required for the amorphization of the $Gd_2Ti_2O_7$ is about 0.143 DPA [51]. The same group had also reported amorphization of the zirconate pyrochlore to be ~5.5 DPA at 300 °K, which is normally considered as radiation resistant. On annealing, recovery rate for radiation-induced defects in fully amorphized structure ($La_2Ti_2O_7$) is found to be higher than amorphous structure ($Gd_2Ti_2O_7$] due to critical role played by the topology of the amorphous materials [52]. This puts La-based composite in better condition than Gd composites. Also, it may be added that the ratio, $\sigma_{el}/\sigma_{inel}$ for neutrons with energy $E > 1$ MeV is several hundred times higher in case of La than Gd. This is because of lesser absorption cross section of neutrons in La than Gd.

Kulriya et al. [53] have investigated temperature-dependent structural stability of the bulk pyrochlore under swift heavy ion irradiation and observed improvement in the radiation resistant behavior on high temperatures. Also, stability of the bulk pyrochlore structure (space group Fd-3 m) under ion bombardment has been shown to be governed by the ratio of the ionic radii of A and B cations (r_A/r_B). The materials like $La_2Zr_2\ O_7$ may further be explored for utilization as structure materials for the reposition and as a layer inside fuel cells.

Other materials that are proposed are SiC and ZrC and PyC in the form of layers in a fuel cell. Radiation resistant behavior of SiC in the spallation neutron spectrum has been found superior when compared with other metals like Ni and Nb. In Sect. 6.4, applications of several other RR materials like Zr_3Si_2 and SiC under the TRIGA setup by the General Atomics has already been discussed.

References

1. Development of Radiation Resistant Reactor Core Structural Materials. https://www.iaea.org/About/Policy/GC/GC51/GC51InfDocuments/English/gc51inf-3-att7_en.pdf
2. Inozemtsev, V., Zeman, A., Scandolo, S.: Development of Radiation Resistant Materials, pp. 20–24, ICTP (April, 2009). http://indico.ictp.it/event/a08149/material/0/0.pdf
3. TM-F1-34567.: Accelerator Simulations and Theoretical Modeling of Radiation Effects Report of the Technical Meeting (IAEA) Hosted by NSC-KIPT, Kharakov, pp. 9–13 (June, 2008). http://www-naweb.iaea.org/napc/physics/meetings/TM34567/login2.html
4. TM-36842.: Physics of Materials Under Neutron and Charged Particle Irradiations, Report of the Technical Meeting Held in Vienna, pp. 16–19, Vienna (November, 2009)
5. Li, M.: Moving from Dpa to Changes in Materials Properties, Talk Delivered at Radiation Effects in Superconducting Magnet Materials (RESMM12), Fermilab, USA 14 Feb 2012. Also see, Frank Garner: Material Science and Technology. https://doi.org/10.1002/9783527603978.mst0110
6. Porollo, S.I.: Influence of silicon on swelling and microstructure in Russian austenitic stainless steel EI-847 irradiated to high neutron doses. J. Nucl. Mater. **378**(1), 17 (2008). http://www-pub.iaea.org/MTCD/Publications/PDF/Pub1548_web.pdf
7. Ryazanov, A.: Comparative physical behaviour of SiC materials under neutron and charged particle irradiations. In: Presented at IAEA Technical Meeting Entitled Physics of Materials Under Neutron and Charged Particle Irradiations, IAEA Head Quarters (November, 2009)
8. Alexander, D.E., Rehn, L.E.: Gamma-ray displacement damage in the pressure vessel of the advanced

boiling water reactor. J. Nucl. Mat. **217**(1994), 213. Also: http://dx.doi.org/10.1016/0022-3115(94)90325-5

9. Cheverton, R.D., Merkle, J.G., Nanstad, R.K.: Oak Ridge National Laboratory Report, ORNL TM-10444 (1988)
10. Cuba, V., Mucka, V., Pospisil, M.: Radiation Induced Corrosion of Nuclear Fuel and Materials. Available from: http://cdn.intechweb.org/pdfs/28495.pdf (Mansur, L.K., Farrell, K.: J. Nucl. Mat., **170**, 236 (1990). http://dx.doi.org/10.1016/0022-3115(90)90294-W)
11. Kislitsin, S.: Investigation of Shape Changes of Hexagonal Ducts of Spent Fuel Assemblies from the BN-350 Fast Neutron Nuclear Reactor TM-36842, Vienna, 16–19 Nov 2009. Reproduced by the International Atomic Energy Agency
12. Wallenius, J.: Radiation Damage & GAS Production in Metals, Materials for Gen IV Nuclear Reactors, MATGEN-IV.2 from Iron to Steel. The Ice Hotel, Jukkasrvi, Sweden, 2–7 Feb 2009
13. Chung, H.M., Loomis, B.A., Smith, D.L.: Development and testing of vanadium alloys for fusion applications. J. Nucl. Mat. **239**, 139 (1996)
14. Dienes, G.J., Vineyard, G.H.: Radiation Effects in Solids. Interscience, New York (1957)
15. Kinchin, G.H., Pease, R.S.: The displacement of atoms in solids by radiation. Rep. Progr. Phys. **18**, 1 (1955)
16. Nelson, R.S.: AERE Report, R-6092 (1969)
17. Robinson, M.T., Torrens, I.M.: Computer simulation of atomic-displacement cascades in solids in the binary-collision approximation. Phys. Rev. B **9**, 5008 (1974)
18. Norgett, M.J., Robinson, M.T., Torrens, I.M.: A proposed method of calculating displacement dose rates. Nucl. Eng. Des. **33**, 50 (1975)
19. Lindhard, J., et al.: Integral equations governing radiation effects. Kgl. Dansk, Vid. Selsk, Mat. Fys. Medd. **33**, 10 (1963), also, gymarkiv.sdu.dk/MFM/kdvs/mfm%2030-39/mfm-33-10.pdfgymarkiv.sdu.dk/MFM/kdvs/mfm%2030-39/mfm-33-10.pdf
20. Gopalakrishnan, V.: Nuclear Data Section, IGCAR, Kalpakkam, India, RPD/NDS/96 (2001)
21. Kumar, V., Raghaw, N.S., Palsania, H.S.: A Monte Carlo Code for Radiation Damage by Neutrons. Nucl. Sci. Eng. **172**, 151 (2012)
22. Tundwal, A., Kumar, V., Raghaw, N.S.: Monte Carlo Simulation of Radiation Damage by Gamma Rays, p. 749. IEEE Xplore (2014). https://doi.org/10.1109/epe.2014.6839457 [and for details, Ambika Tundwal: Ph.D. thesis entitled: A Monte Carlo Code for Radiation Damage of Reactor Fuel cells, Steel and Polyethylene by High Energy Gamma and Electrons, April (2016), GGSIP University, New Delhi (India)]
23. Phythian, W.J.: A comparison of displacement cascades in copper and iron by molecular dynamics and its application to microstructural evolution. J. Nucl. Mat. **223**(3), 245 (1995). https://doi.org/10.1016/0022-3115(95)00022-4
24. Stroller, R.F.: Point defect survival and clustering fractions obtained from molecular dynamics simulations of high energy cascades, J. Nucl. Mater. **1996**, 233–237, 999 (1996). http://dx.doi.org/10.1016/S0022-3115(96)00261-9
25. Alonso, E., Catrula, M.J., Diaz De La Rubia, T., Perlado, J.M.: Simulation of damage production and accumulation in vanadium. J. Nucl. Mat. **276**, 221 (2000). Accessed http://dx.doi.org/, https://doi.org/10.1016/s0022-3115(99)00181-6
26. Robinson, M.T.: The binary collision approximation: Background and introduction. In: Proceedings of the International Conference on Computer Simulations of Radiation Effects in Solid, Hahn-Meitner Institute Berlin, Berlin, Germany, 23–28 Aug (1992)
27. Ziegler, J.F., Biersack, J.P., Littmark, U.: The Stopping and Range of Ions in Solids. The Stopping and Ranges of Ions in Matter, Vol. 1. Pergamon Press, New York (1985)
28. Ziegler, J.F., Biersack, J.P.: Updated Version of a Computer Code for Calculating Stopping and Ranges. The Stopping and Ranges of Ions in Matter, Vol. 1. Pergamon Press, New York (1985)
29. Robinson, M.T.: Slowing-down time of energetic atoms in solids. Phys. Rev. B **40**, 10717 (1989)
30. MARLOWE (Version 13): Available from the Radiation Shielding Information Center, Oak Ridge National Laboratory, Post Office Box 2008, Oak Ridge, Tennessee 37831-6362, U.S.A.
31. Srinivasan, P.: Monte Carlo Simulation of Neutron Phenomena, Applications of Monte Carlo Methods in Nuclear Science and Engineering. ISBN: 978-81-8372-047-2 (April, 2009)
32. Broeders, C.H.M., Yu, A., K. Konobeyev, Voukelatou, K.: IOTA—A Code to Study Ion Transport and Radiation Damage in Composite Materials. Forschungszentrum Karlsruhe GmbH, FZKA 6984, Karlsruhe, Germany, ISSN: 0947-8620 (2004)
33. ASTM E521-96.: Standard Practice for Neutron Radiation Damage Simulation by Charged Particle Radiation, Annual Book of ASTM Standard, vol. 12.02. American Society of Testing and Material, Wes Conshohocken, PA (2009)
34. Zarkadoula, E.: The nature of high-energy radiation damage in iron. J. Phys. Condens. Matter. **25**, 125402 (2013)
35. International Atomic Energy Agency. Evaluated Nuclear Data File-ENDF B VII.0. https://www-nds.iaea.org/public/download-endf/. Accessed 18 Jan 2009
36. Mayol, R., Salvat, F.: Atomic Data and Nuclear Data Tables, pp. 55–154 (1997). http://dx.doi.org/10.1006/adnd.1997.0734 (Article No. DT970734, 65 1)
37. Raghaw, N.S., et al.: A monte carlo simulation of radiation damage of SiC and Nb using JA-IPU code. J. Energy Power Eng. **9**, 967 (2015)
38. Kumar, V., Raghaw, N.S., Tundwal, A., Korovin, Y., Adam, J.: A Study of Radiation Resistant Material using JA-IPU Code of Radiation Damage: A Talk Presented at the Annual Scientific session

NRNU-MEPh-2015 held at INPE, Obninsk from 1–4 Feb 2015
39. Tundwal, A., Kumar, V., Datta, A.: Gamma radiation induced resistivity changes in iron. Ind. J. Phys. **91**, 293 (2017)
40. Brinkman, J.A.: Production of atomic displacements by high energy particles. Am. J. Phys. **24**(4), 246 (1956). http://dx.doi.org/10.1119/1.1934201
41. Kumar, V., Tundwal, A., Vijay, Y.K., Palsania, H.S.: Investigation of defect production in iron on gamma irradiation using the doppler broadening technique. Ind. J. Pure App. Phys. **54**, 51 (2016)
42. Was, G.S.: Fundamental of Radiation Material Science: Metal and Alloys. Springer, Berlin, Heidelberg, New York. ISBN 978-3-540-49471-3 (2007)
43. Voter, A.F.: Introduction to the Kinetic Monte Carlo Method. In: Radiation Effects in Solids. Springer (2005). Also see indico.ictp.it/event/a09140/session/28/contribution/19/material/0/0.pdf
44. Tundwal, A., Kumar, V., Raghaw, N.S., Datta, A.: Monte carlo simulation of radiation damage produced in iron and vanadium by primary knock on atom 'PKA'. Radiat. Eff. Defects Solids **171**(7–8), 658 (2016)
45. Raghaw, N.S.: Ph.D. Thesis Entitled: A Study of Radiation Damage of Materials Irradiated by High Energy Neutron and Charged Particles—A Simulation and Validation, (June 2016), University of Rajasthan, Jaipur, India
46. Alexander, D.E., Rehn, L.E.: The contribution of high energy gamma rays to displacement damage in LWR pressure vessels. J. Nucl. Mat. **209**, 212 (1994)
47. Tundwal, A., Kumar, V.: Estimation of radiation damage of iron by a reactor gamma spectrum. Kerntechnik **80**(5) (2015)
48. Ewing, R.C., et al.: Nuclear waste form for the immobilization of plutonium and "Minor" actinides. J. Appl. Phys. **95**, 5949 (2004)
49. Burns, P.C., et al.: Nuclear fuel in a reactor accident. Science **335**, 1184 (2012) and Ewing R.C. Long-term storage of spent nuclear fuel. Nature Materials **14**, 252 (2015)
50. Lian, J., et al.: Radiation-induced effects in pyrochlores and nanoscale materials engineering. Nucl. Instr. Methods Phys. Res. B **250**, 128 (2006)
51. Lian, J., et al.: Ion-beam implantation and cross-sectional TEM characterization of Gd2Ti2O7 pyrochlore. Nucl. Instr Methods Phys. Res B **242**, 448 (2006)
52. Park, S., et al.: Response of $Gd_2Ti_2O_7$ and $La_2Ti_2O_7$ to swift ion irradiation and annealing. Acta Mater. **93**, 1 (2015)
53. Kulriya, P.K., et al.: In-situ high temperature irradiation setup for temperature dependent structural studies of materials under swift heavy ion irradiation. Nucl. Instr. Methods in Phys. Res. B **342**, 98 (2015)

CPSIA information can be obtained
at www.ICGtesting.com
Printed in the USA
LVHW060040171218
600703LV00011B/795/P

9 789811 075025